T0211924

Ending Medicine's Chronic Dysfunction: Tools and Standards for Medical Decision Making

Synthesis Lectures on Assistive, Rehabilitative, and Health-Preserving Technologies

Editors

Ronald M. Baecker, *University of Toronto*
Andrew Sixsmith, *Simon Fraser University and AGE-WELL NCE*

This series provides state-of-the-art overview lectures on assistive technologies. We take a broad view of this expanding field, defining it as information and communications technologies used in diagnosis and treatment, prosthetics that compensate for impaired capabilities, methods for rehabilitating or restoring function, and protective interventions that enable individuals to stay healthy for longer periods of time.

Each overview introduces the medical context in which technology is used, presents and explains the technology; reviews problems and opportunities, successes and failures in the development and use of technology; and synthesizes promising opportunities for future progress. Authors include significant material based on their own work, while surveying the broad landscape of an area's research, development, and deployment progress and success.

Ending Medicine's Chronic Dysfunction: Tools and Standards for Medical Decision Making
Lawrence L. Weed (1923–2017) and Lincoln Weed

Research Advances in ADHD and Technology
Franceli L. Cibrian, Gillian R. Hayes, and Kimberley D. Lakes

AgeTech, Cognitive Health, and Dementia
Andrew Sixsmith, Judith Sixsmith, Mei Lan Fang, and Becky Horst

Interactive Technologies and Autism, Second Edition
Julie A. Kientz, Gillian R. Hayes, Matthew S. Goodwin, Mirko Gelsomini, and Gregory D. Abowd

Zero-Effort Technologies: Considerations, Challenges, and Use in Health, Wellness, and Rehabiliation, Second Edition
Jennifer Boger, Victoria Young, Jesse Hoey, Tizneem Jiancaro, and Alex Mihailidis

To Margo (1952-2014) and Julia

Ending Medicine's Chronic Dysfunction: Tools and Standards for Medical Decision Making
Lawrence L. Weed (1923–2017) and Lincoln Weed

ISBN: 978-3-031-00479-7   print
ISBN: 978-3-031-01607-3   ebook
ISBN: 978-3-031-00041-6   hardcover

DOI 10.1007/978-3-031-01607-3

A Publication in the Springer series
SYNTHESIS LECTURES ON ASSISTIVE, REHABILITATIVE, AND HEALTH-PRESERVING TECHNOLOGIES
Lecture #16

Series Editors: Ron Baecker, University of Toronto and Andrew Sixsmith, Simon Fraser University and AGE-WELL NCE

Series ISSN 2162-7258 Print   2162-7266 Electronic

Ending Medicine's Chronic Dysfunction: Tools and Standards for Medical Decision Making

Lawrence L. Weed (1923–2017)

Lincoln Weed

SYNTHESIS LECTURES ON ASSISTIVE, REHABILITATIVE, AND HEALTH-PRESERVING TECHNOLOGIES #16

ABSTRACT

This book describes an overlooked solution to a long-standing problem in health care. The problem is an informational supply chain that is unnecessarily dependent on the minds of doctors for assembling patient data and medical knowledge in clinical decision making. That supply chain function is more than the human mind can deliver. Yet, dependence on the mind is built into the traditional role of doctors, who are educated and licensed to rely heavily on personal knowledge and judgment. The culture of medicine has long been in denial of this problem, even now that health information technology is increasingly used, and even as artificial intelligence (AI) tools are emerging. AI will play an important role, but it is not a solution. The solution instead begins with traditional software techniques designed to integrate novel functionality for clinical decision support and electronic health record (EHR) tools. That functionality implements high standards of care for managing health information. This book describes that functionality in some detail. This description is intended in part to be a starting point for developers in the open source software community, who have an opportunity to begin developing an integrated, cloud-based version of the tools described, working with interested clinicians, patients, and others. The tools grew out of work beginning more than six decades ago, when this book's lead author (deceased) originated problem lists and structured notes in medical records. The electronic tools he later developed led him to reconceive education and licensure for doctors and other health professionals, which are also part of the solution this book describes.

KEYWORDS

problem-oriented, electronic health records, clinical decision support systems, artificial intelligence, clinician role, licensure

Contents

CHAPTER 1

Introduction

… the avoidance of reality is much the same everywhere… To see what is in front of one's nose needs a constant struggle.— George Orwell[1]

1.1 WHAT IS THIS BOOK ABOUT?

Soon after outbreak of the COVID-19 pandemic in the U.S., the supply chain for essential medical equipment reached a breaking point.[2] Medicine has yet to recognize a similar, older, and larger problem: constant failures in the supply chain for health information. Unlike equipment for a pandemic, health information cannot be mass produced in advance. Patient care requires constant assembly of information, both knowledge and data. That information must be selected, distilled, and organized for patient-specific, problem-specific use in every case. As the assembled information is used for decision making, maintaining the supply chain requires meticulous, organized recordkeeping and follow-up over time.

Constant failures in these supply chain activities compromise decision making across the board. And quality failures occur with not only medical decisions (missed diagnoses or misguided treatments, for example) but execution of decisions (unskilled physical examinations or surgeries, for example). Taken together, these two areas of quality failure—decision making and execution—have long caused harm and economic waste on a global scale. Continuing to tolerate these quality failures is comparable to inaction against a disease pandemic.

This book focuses on medical decision making, not execution ("doing the right thing," not "doing the thing right").[3] Decision making depends on medicine's informational supply chain. So, the book's core subject matter is to diagnose and remedy failures in that supply chain.

---

[1] "In Front of Your Nose" (1946), in Orwell, G. *Narrative Essays* (Penguin Books, 2009).

[2] See Mukherjee, S., "What the Coronavirus Reveals About American Medicine," *The New Yorker*, April 27, 2020. Dr. Mukherjee's article goes on to discuss electronic health records (EHRs). See Section 3.4. Dr. Mukherjee concluded that once the pandemic is over, we should do more than merely resume the status quo. "We need to think not about resumption but about revision." This book argues that we need to think not about revision but about reconstruction from the foundation.

[3] Decision making has also been referred to as planning. See the famous report by the Institute of Medicine (now the National Academy of Medicine or NAM), *To Err Is Human: Building a Safer Health Care System* (National Academies Press, 2000), p. 55, which explains: "This report addresses primarily … errors of execution, since they have their own epidemiology, causes and remedies that are different from errors of planning. Subsequent reports … will address the full range of quality related issues, sometimes classified as overuse, underuse and misuse"). The first of the subsequent reports was *Crossing the Quality Chasm: A New Health System for the 21st Century* (National Academies Press, 2001), which repeatedly referenced the work of this book's lead author, Dr. Lawrence L. Weed (referred to below as LLW). For background on LLW, see Section 1.3 and Chapter 5.

Developing an effective informational supply chain is a pathway for improving the quality of medical decisions. Improved decisions are needed for improving the economics of the health domain. And improved economics is needed to afford universal coverage without diverting resources from other urgent needs. In short, everyone in the health domain has a large stake in using this new pathway.

The root cause of supply chain failures is misguided dependence on the human mind. Yet the role of the human mind has long been venerated. The culture of medicine is thus in denial about this root cause.

The state of denial makes it difficult to recognize and remedy dependence on the mind, even now as digital information tools increasingly offer an escape. That misguided dependence, and the continuing state of denial about that problem, are the subject of Part I of this book. Solving that problem requires new digital tools for health records and decision support based on new standards of care for handling health information—the subject of Part II.[4] New tools and standards make possible new occupational roles for doctors and other health professionals, new concepts of medical education and licensure, and new concepts of expertise for both health professionals and their patients—also the subject of Part II.

Much of the critique in Part I will sound familiar to readers engaged in health policy issues. Decades ago, this book's lead author (LLW) was among the first to recognize issues like the centrality of medical records, the need to protect against, rather than venerate, autonomous expert judgment, and the primacy of the patient's role. Now that these insights have become conventional wisdom, the question arises—could there be something more in LLW's work that we're still missing? That something more is the solution presented in Part II, which has never been widely understood, much less adopted.[5]

---

[4] We addressed both the problem and the solution in our book, *Medicine in Denial* (Amazon Kindle Direct Publishing, 2011). The present book draws heavily on *Medicine in Denial*. Its full text is available for free at www. world3medicine.org as a downloadable, searchable PDF. See also a thoughtful series of blog posts on *Medicine in Denial* by Dr. Leslie Kernisan (the third and last post with links to the earlier ones is available here). Readers may also be interested in Weed, L. L. et al., *Knowledge Coupling: New Premises and New Tools for Medical Care and Education* (Springer-Verlag, 1991). Lincoln Weed has dozens of copies of that book, available for free. Contact ldweed424@gmail.com or text to 703-424-4408. The present book covers only a fraction of the material covered by the *Medicine in Denial* and *Knowledge Coupling* volumes.

[5] Some readers may ask how this book relates to the sweeping critiques of the medical-industrial-academic complex presented in Ivan Illich's prophetic *Medical Nemesis: The Expropriation of Health* (Pantheon Books, 1976) and in Seamus O'Mahony's recent *Can Medicine Be Cured? The Corruption of a Profession* (Apollo, 2019). (See this blog post discussing both books: Richard Smith: The most devastating critique of medicine since Medical Nemesis by Ivan Illich in 1975 (February 13, 2019).) LLW, who read the Illich book, sought to develop solutions to the failings he saw in real-world problem-solving for patients. His pursuit of those solutions (not mere analysis of failings) led him to the diagnosis stated above—misguided reliance on the minds of doctors—which is a primary root cause of both patient care and system-wide failings. The human mind is the vector for the systemic corruptions in the medical-industrial-academic complex as described by Illich and O'Mahony. Unlike them, LLW focused on solutions, which deepened his insight into the failings. LLW's solutions represent a way to harvest value for patients and society from the medical-industrial-academic complex while guarding against its exploitive tendencies.

The following case vividly illustrates the problem described in Part I, and it points straight at the solution in Part II.

1.1.1   THE CASE OF THE CRAZY SURGEON AND THE WISE PHYSICAL THERAPIST

The patient, Eric, had an unusual disease in his joints—osteochondritis dissecans. It caused Eric's knees to go bad during his teens and by age 20 he had undergone reparative surgery on both knees.

Over the next 40 years, he had to progressively curtail his physical activities as his repaired knees deteriorated. At age 62 he decided to have replacement surgery on one knee. He relied on advice from his orthopedic surgeon that he was a perfect candidate for knee replacement. The orthopedist said the only significant downside was a 1–2% risk of infection.

Two days after the surgery, Eric began the standard post-operative physical therapy (PT) protocol. But immediately something was wrong. As described by Eric:

> The protocol is intense, calling for aggressive bending and extension to avoid scar formation in the joint. Unable to get meaningful flexion, I put a stationary bicycle seat up high and had to scream in agony to get through the first few pedal revolutions. The pain was well beyond the reach of oxycodone. A month later the knee was purple, very swollen, profoundly stiff, and unbending. It hurt so bad that I couldn't sleep more than an hour at a time, and I had frequent crying spells.

Eric's physical therapist had no solution, and his orthopedist dismissed the pain. "You should have your internist prescribe anti-depression medications," the orthopedist said. Eric and his wife listened "in total disbelief" to this "robotic response." Eric goes on:

> That seemed crazy enough. But the surgeon then recommended a more intensive protocol of physical therapy, despite the fact that each session was making me worse. I could barely walk out of the facility or get in my car to drive home. The horrible pain, swelling, and stiffness, were unremitting. I became desperate for relief, trying everything … fully aware that none of these putative treatments have any published data to support their use.

Then Eric's wife learned about arthrofibrosis—a disastrous complication suffered by 2–3% of patients having knee replacement surgery, which is the most common of all orthopedic operations. Arthrofibrosis causes a vicious inflammation response and profound scarring. Eric confirmed with his orthopedist that he had this condition. The orthopedist told him there was nothing to be done but wait for a year, at which point more surgery could remove the scar tissue. "The thought of going a year as I was or having another operation," Eric wrote, "made me feel even sicker."

Then a friend recommended a different physical therapist. She rescued Eric from his nightmare:

Over the course of 40 years, she had seen many patients with osteochondritis dissecans, and she knew that, for patients such as me, the routine physical therapy protocol was the worst thing possible. … her approach was to go gently; she had me stop
all the weights and exercises and use anti-inflammatory medications. She handwrote
a page of instructions and texted me every other day to ask how "our knee" was doing.
Rescued, I was quickly on the road to recovery. Now, years later, I still have to wrap
my knee every day to deal with its poor healing. So much of this torment could have
been prevented.

Eric believes that "a full literature review … might well have indicated that I needed a special, bespoke PT protocol," assuming that "experienced physical therapists such as the woman I
eventually found shared their data" via the medical literature. He also suggests that the information
could have been shared via artificial intelligence, usable by not only his clinicians but him as a patient. But no such tool was available. In its absence, the orthopedist needed to review the literature,
or consult someone else, or otherwise research the correct therapy for a patient like Eric. But he
evidently failed to do so:

As it was, I was blindsided, and *my orthopedist hadn't even taken my history of osteochondritis dissecans into account* when discussing the risk of surgery, even though he
later acknowledged that it had, in fact, played a pivotal role in the serious problems
I encountered [emphasis added].

Eric goes on to discuss the roles played by his orthopedist and the second physical therapist.
Referring to the follow-up visit where the orthopedist had suggested an anti-depressant, Eric writes:

The idea that I should take medication for depression exemplifies a profound lack
of human connection and empathy in medicine today. Of course, I was emotionally
depressed, but depression wasn't the problem at all: the problem was that I was in
severe pain and had Tin Man immobility. The orthopedist's lack of compassion was
palpable: in all the months after the surgery, he never contacted me once to see how
I was getting along. The physical therapist not only had the medical knowledge and
experience to match my condition, but she really cared about me. It's no wonder that
we have an opioid epidemic when it's a lot quicker and easier for doctors to prescribe
narcotics than to listen to and understand patients.[6]

Tragically, Eric's case is not an isolated occurrence. It is all too representative of decision
making that is uninformed and callous. Eric writes:

*Almost anyone with chronic medical conditions has been "roughed up" like I was—it hap-
pens all too frequently … the problem is so pervasive that even insider knowledge isn't
necessarily enough to guarantee good care.* Artificial intelligence alone isn't going to

---

[6] We return to the issue of opioid prescribing in discussing medicine's lack of generally accepted standards of care
for managing health information. See Sections 3.2.1 and 9.1.3.3.

solve this problem on its own. We need humans to kick in. *As machines get smarter and take on suitable tasks, humans might find it actually easier to be more humane.* [Emphasis added.]

Note the mention of "insider knowledge." Eric is an "insider" because he is Eric Topol, M.D., a distinguished cardiologist and the author of three books about the future of medicine. The above description of his case is paraphrased and quoted from the opening pages of his most recent book, *Deep Medicine: How Artificial Intelligence Can Make Healthcare Human Again.*[7]

1.1.2 HOW THIS CASE EXEMPLIFIES THE SUBJECT MATTER OF THIS BOOK

What happened to Dr. Topol reveals how quality failures are built into medical practice at multiple levels. Specifically:

- The orthopedist caused four harmful breakdowns in the informational supply chain. (1) He failed to inform his patient of the general arthrofibrosis risk and the special risk arising from a history of osteochondritis dissecans. (2) Once that special risk materialized, he failed to diagnose it. (3) That diagnostic failure led him to recommend an even harsher PT regimen and to suggest an anti-depressant. (4) These errors, together with his lack of compassion, completely alienated his patient.

- That alienation is itself a breakdown in the informational supply chain, because close communication between patient and clinician is essential to quality of care. Sometimes alienation causes patients to distrust the entire health care system. Their distrust may then deter them from seeking care when they need it—the ultimate supply chain breakdown. And, like a metastasizing cancer, the patient's distrust may spread to family members and others who witness what happened.

- Dr. Topol's suffering and disability came to an end only because a friend happened to recommend a physical therapist whose knowledge and experience happened to match his particular needs. This random matching falls far short of what we should expect from a scientifically rigorous system of care. Yet, randomness is inevitable if medicine's informational supply chain depends on the clinician's mind, with all its fallibility and idiosyncrasies. Medicine cannot be practiced as a *science*, and care cannot be provided through a *system*, until the supply chain incorporates tools and standards external to any clinician's mind. Clinicians must be governed by standards of care assuring that the tools are used *habitually* from the outset of care to match medical knowledge with

---

[7] *Basic Books* (2019), pp. 1-4. The prior two books are *The Patient Will See You Now* (Basic Books, 2015) and *The Creative Destruction of Medicine* (Basic Books, 2012).

detailed patient data carefully selected by the tools. The minds of clinicians and patients still have a supplemental role to play, but they cannot trust their own minds to perform this matching process.

- The needed information tools, Dr. Topol suggests, should take the form of artificial intelligence (AI) software. He believes that AI "could have predicted that my experience after the surgery would be complicated" (p. 3). But advanced AI software is not necessary for a case like Dr. Topol's. Traditional software optimized for general medical decision making, as described in Chapters 8 and 9, could have been used to identify the arthrofibrosis complication risk, the effect of the osteochondritis dissecans history, and the alternative to the standard PT protocol. This point is crucial, because traditional software already offers simplicity, reliability, and transparency that are not yet feasible with advanced AI in many contexts. Moreover, traditional software offers protections against some risks inherent in both human and artificial intelligence.

- External tools are not enough. For both traditional software and AI tools, their design and use must be governed by high standards of care for managing complex health information. The key point is that external tools make it possible to set standards of care at a higher level than the unaided human mind can attain. Hopelessly complex knowledge and data become manageable when we use external tools designed to implement high standards of care for managing health information.

- "Standards of care" are not technical standards (e.g., data, vocabulary, and interoperability specifications, which is what "standards" suggests to health IT specialists). Instead, standards of care are rules of conduct. For example, Dr. Topol's orthopedist violated a rule of conduct that clinicians must *habitually* employ external tools to match general knowledge with patient-specific data. Another rule of conduct is maintaining accurate, organized records in compliance with generally accepted standards for health recordkeeping, not unlike generally accepted accounting principles for financial records. Complying with these standards of care is a matter of both scientific integrity and public safety, not to mention efficiency and productivity.

- The rules of conduct just described are process standards. Also needed are detailed informational standards, specifying the knowledge and data that must be taken into account for a given problem situation. For Dr. Topol, the initial problem situation was deciding whether to undergo knee replacement surgery and rehab. A good informational supply chain would have prompted a question on whether a candidate for knee replacement surgery has a history of osteochondritis dissecans. (Ideally, the question would be answerable by automated search of a lifetime electronic health record, without the need for the patient to recall that history.) Given a positive response to that question, the informational supply chain would inform the patient and clinician of the arthrofibrosis risk and the need for a special PT regimen. In other words, everyone involved should not have had to rely on their own minds or the minds of others. Their personal ignorance or awareness should not have mattered.

- A good informational supply chain protects against conflicts of interest. Maybe Dr. Topol's orthopedist deliberately understated the risks of knee replacement surgery to induce his patient to elect surgery and thus generate fees. Or suppose an insurer denied coverage for the surgery or the post-operative PT on the ground that neither would work in a patient with a history of osteochondritis dissecans. Or suppose Dr. Topol never learned of the physical therapist with the requisite knowledge. Safeguards against all these random contingencies are provided by the right tools and standards. Everyone involved should have been able to consult an external tool supplying the information needed.

- Clinician roles are pivotal. The informational supply chain function is so deeply embedded in the doctor's role that Dr. Topol trusted his orthopedist's advice. Yet, the second physical therapist, not the orthopedist and not the first physical therapist, turned out to have the expertise relevant to Dr. Topol's needs. Why is it, then, that an orthopedist is higher in the medical hierarchy than a physical therapist? This disparity in status further manifests randomness and dysfunction in the informational supply chain. And it calls into question current notions of licensure and authority for doctors and all medical practitioners.

- It may seem surprising that Dr. Topol, as an "insider," trusted his surgeon's advice. But this is understandable. No one can easily find the time to second guess professional advice and conduct independent research. As a specialist himself, Dr. Topol would have known how time-consuming it might become for him to delve into a different specialty, orthopedics. And he likely trusted a fellow doctor to be as diligent and caring as he would be. So the path of least resistance was to trust the advice he received. But the reality of medicine is that such trust is too often not warranted, from both medical and economic perspectives.

- From a medical perspective, even if doctors were always diligent and caring, they are attempting the impossible. Their minds are not capable of functioning reliably as part of the informational supply chain. Were the human mind treated as a medical device for performing this function, it would never pass FDA scrutiny for safety and effectiveness.

- From an economic perspective, trusting doctor advice may seem rational, particularly with specialists. Specialization normally fosters expertise and efficiency. But this principle doesn't apply if the specialist's expertise doesn't match patient needs. In medicine that mismatch is the norm, because patient needs routinely cross specialty boundaries. This point implicates medicine's division of labor. External tools can take into account all specialties. And that expertise can be built into the tools *before* patient encounters, obviating the need for each clinician to reinvent the wheel *during* those encounters. The tools can be used by all clinicians, not just doctors, and by patients themselves, before, during, and after their encounters. An economically rational division of labor among external tools, clinicians, and patients becomes possible once the tools are available. By comparison, the current division of labor in medicine is primitive. Other fields in the domains of science and commerce have evolved away from reliance on opaque expert judgments. It is long past time for medicine to catch up.

- A better informational supply chain would relieve clinicians from the intolerable burden of constantly reinventing the wheel—a burden that contributes enormously to clinician burnout as well as negligent and callous conduct such as that of Dr. Topol's orthopedist. Dr. Topol recognized this reality: "As machines get smarter and take on suitable tasks, humans might find it actually easier to be more humane."[8]

The above points detail various ways that Dr. Topol's case exemplifies the subject matter of this book. The following section further introduces the book's core concept: that medicine's failures of quality and economy are rooted in the doctor's role.

1.1.3   A NEW DIVISION OF LABOR

Discontent with the status quo and ongoing attempts at reform have long been part of the culture of medicine. Yet that culture has never faced up to how the doctor's role blocks true reform. What that means, and what to do about it, are the subject of this book.

Fundamentally changing the doctor's role is an alternative outside mainstream health reform. This alternative, despite having evolved for six decades, is genuinely new—it has not truly been considered, much less accepted or advocated, by mainstream experts. Nor have they rejected this alternative; it's not even on their radar screens. That is the case even though mainstream economists have long criticized the doctor's role. Locked in place by medical licensure, the doctor's role,

---

[8] *Deep Medicine* (see Note 7), p. 4. Dr. Topol later illustrates this point with a quote from Lynda Chin: "Imagine if a doctor can get all the information she needs about a person in 2 minutes and then spend the next 13 minutes of a 15 minute office visit talking with the patient, instead of spending 13 minutes looking for information and 2 minutes talking with the patient." Ibid., p. 23.

according to this critique, is a monopolistic barrier to market forces that would otherwise generate major innovations in medical practice.[9]

We tend to agree with the economists' critique as far as it goes. But it doesn't go very far, because it doesn't envision an alternative to the doctor's role. The critique needs to be reconciled with economists' traditional view of the doctor's role as a regulatory necessity.[10] Because that traditional view persists, mainstream health policy thinking still accepts the doctor's role as inherent in modern medicine. It involves advanced science and technology, so everyone is socialized to believe that medicine must be practiced by, or under the direction of, highly educated, highly compensated doctors. They alone are thought to have the trained minds needed to apply medical science. Their training is thought to confer some sort of unified expertise on doctors as an occupational category.

The alternative view advocated in this book is that the doctor's role is unnecessary to—indeed *incompatible with*—scientific rigor in medical practice. Without scientific rigor, medicine will continue to be practiced as an artisanal craft or a commercial endeavor, both at widely varying levels of skill and quality, both poorly connected to patient needs, and both far less productive and affordable than they must become.

So what exactly is the doctor's role, and how is it an obstacle to scientific rigor? We address those questions in Chapter 2 and give detailed examples in Chapter 3.

Finally, we need to explain how this book's subject matter fits into the Morgan & Claypool "Synthesis Lectures on Assistive, Rehabilitative, and Health-Preserving Technologies." The series concerns "the expanding field of assistive technologies," defined broadly to include technologies in prosthetics, rehabilitation, and protective interventions to preserve health. The subject matter encompasses LLW's argument that *everyone*, patients and clinicians alike, must *habitually* use "assistive" information tools as a cognitive "prosthetic" to "rehabilitate" the human mind from its innate limits. This argument goes well beyond the current consensus that cognitive heuristics and biases impair medical decision making. Moreover, with recent pathbreaking advances in AI, a related issue is emerging: in what contexts should advanced AI play more than an "assistive" role? When should it supersede human intelligence?

---

[9]   See *Medicine in Denial* (Note 4), part VIII.B (pp. 210–219), and Section 2.1. The reference to "market forces" should not be taken as endorsing an unregulated "free market" in health care. On the contrary, the potential of market forces cannot be realized unless medical practice is tightly regulated in very specific ways. Those specific ways were the subject of the lead author's work for six decades.

[10]   This traditional view was articulated by the late Kenneth Arrow in a seminal 1963 article, which we have discussed elsewhere. See Note 121.

1.2   WHO IS THE AUDIENCE?

Without the necessary information tools, the reforms envisioned in this book are powerless to be born. So this book's most immediate audience, though not the only one, is those who are best positioned to start building the necessary tools—developers in the open source software community. This is not limited to specialists in health IT. Open source developers outside health IT may have fresh perspectives and relevant experience.

Health IT differs somewhat from other fields where software developers work. Developers normally lack expertise in the functions to be performed by the tools they build. So they normally rely on guidance from subject matter experts in those functions. The authoritative subject matter experts in medicine are doctors.

Yet, this book is in part a critique of the doctor's role, including mainstream doctor practices and orthodoxies in medical decision making. If our critique is right, then the doctor's role poses a dilemma for software developers. Their dilemma is that they must combine a healthy respect for the firsthand clinical experience of doctors with a healthy skepticism of guidance from doctors on designing digital tools for clinical use. In other words, developers cannot rely on guidance from doctors in the way they would normally rely on guidance from subject matter experts.

Doctors are not the only subject matter experts who health IT developers work with. Others are non-doctor clinicians, business/administrative managers, academic specialists, and regulators (not to mention patients, who have personal expertise, as we discuss elsewhere). But, like doctors, all these experts have perspectives that do not always work as guidance to health IT developers. Moreover, the agendas of medical and non-medical experts may come into conflict.

A prime example of conflict is electronic health record (EHR) tools. Clinicians rightly believe that EHRs have been subverted by an economic agenda—generating documentation to maximize billing and reimbursement. The result is that doctors bitterly criticize EHRs as hard-to-use, clinically dysfunctional, and thus extraordinarily burdensome. This mismatch between clinical and economic agendas in EHR design is indeed a serious problem, so much so that cumbersome EHRs are a major contributor to clinician burnout.[11] This is one of the mainstream failings that the open source software community can help address.

All this means that software developers should not accept on faith the directions or guidance they receive from clinicians and other subject matter experts in health care. The open source software community, indeed everyone in the health domain, should not uncritically buy into either mainstream orthodoxy or the contrary views stated in this book. Instead, everyone can independently think through how the health care system can be reformed, taking into account their

---

[11] See Section 3.4.4, which cites recent articles written for the general public. Countless more descriptions can be found in the medical literature and blog posts.

personal experiences in the health domain and in their own professions. Unlike everyone else, however, software developers are positioned to bring to life essential information tools.

Moreover, developers have special expertise in what it takes to translate work processes of many kinds into step-by-step procedures capable of being executed by traditional software tools. They also have expertise in structuring database repositories in the cloud for use by various parties in disparate contexts. And they are developing expertise in advanced artificial intelligence, including machine learning. Those kinds of expertise are highly relevant.

In short, we hope this book elicits interest among software developers in building essential information tools on an open source basis. Some of those readers may wish to jump ahead directly to Part II of this book. The portions most directly relevant are Section 8.3 (discussing a seven-step sequence to be implemented in clinical decision support tools), Section 9.1.2 (discussing medical record structure and data categories), Section 9.2 (discussing a consolidated record in the cloud with a common data model for patient data and care processes), and Section 9.3 (discussing one company's important work on integrating clinical decision support with electronic health records). That said, the entire book is relevant to software developers, who need some understanding of the clinical and policy issues involved.

Although the software developer audience is primary, that is certainly not the only audience. Building and using the tools ultimately requires a common understanding among developers, clinician and patient users of what they develop, institutional providers, third-party payers, regulators, researchers, health policy analysts, and other stakeholders in the health domain, regardless of political orientation. They all can think through the ideas presented, and seek to arrive at some common understanding. In particular, they can think through what the essential information tools involve, in terms of purpose, design, proper and improper uses, limitations, and potential effects, in relation to existing health IT. Most of all, individual consumers/patients and health professionals need to understand how they can jointly use the tools to meet individual health needs.

Accordingly, this book is written for a broad audience. That means the book includes some basic material unnecessary for some readers—for example, basic concepts familiar to software developers in the health IT field and basic medical concepts familiar to clinicians. It also means the book is intended to be accessible to readers from a wide range of educational levels and occupational backgrounds.

1.3   WHO ARE THE AUTHORS?

The lead author, Dr. Lawrence L. Weed (referred to below as LLW), died in 2017.[12] Although he did not participate in the actual writing of this book, he is named as the lead author for two reasons.

---

[12]   See his *New York Times* obituary, this tribute ("Remembering Larry Weed") by Art Papier in *MedPage Today*, this tribute (and the links therein) by Dr. Papier to the Society for Improving Diagnosis in Medicine, this tribute by Harlan Krumholz in the *Health Affairs* blog, and this 2005 article in *The Economist*.

First, LLW originated all the core ideas, spearheaded development and use of the tools and standards described, and inspired immense work by many others over more than six decades. Second, this book draws heavily on the 2011 book *Medicine in Denial* (see Note 4), of which LLW was the lead author. See Chapter 5 for further discussion of LLW's background and influence.

Lincoln Weed is a son of LLW. He is a retired lawyer who, for most of his career, specialized in employee benefits, including health benefits, working initially at a federal agency, and then at law firms in Washington, D.C., and finally at a consulting firm where he focused on health privacy issues for several years. His health benefits work intersected with LLW's work, which led to co-authoring some publications with LLW beginning in 1994, including the *Medicine in Denial* book.

Disclosure: Lincoln and his four siblings inherited LLW's small financial interest (of highly uncertain value) in the buyer of LLW's company, PKC Corp. In addition, Lincoln has a very small interest in another company affiliated with a company developing a decision support and medical record platform related to LLW's work (see Section 9.3). In addition, Lincoln became a member of the non-profit Health Record Banking Alliance (HRBA) and started working with an HRBA committee in early 2020. See Section 9.2 (no compensation is received for working with the HRBA). In addition, LLW and Lincoln have long known the CEO of the company discussed in Section 8.4 but have no financial interest, direct or indirect, in that company.

1.4   BASIC CONCEPTS AND TERMINOLOGY USED IN THIS BOOK

Health and **medical/clinical:** We generally use the adjectives *medical* and *clinical* in reference to the health care system, and the term *health* (as an adjective) in reference to the health domain.

Health care system and **health domain:** We use the term *health care system* to refer to the combined social systems though which health care professionals provide care to patients, professional services are paid for, health care activities are regulated, health-related research is conducted, etc. The health care system is a subset of the health domain. The health care system influences, but does not govern or control, the entire *health domain*.

Health information, **medical/health knowledge**, and **patient data:** *Health information* as an umbrella term covering both general knowledge and patient-specific data. This binary distinction between knowledge and data is useful in many contexts, but is ultimately an oversimplification, as discussed in Section 3.5.

Patient: People act in a *patient* capacity when receiving care within the health care system, and they act in a personal capacity when otherwise engaging in behaviors and pursuits outside the health care system that affect their health. For convenience, however, we sometimes use the term "patient" loosely without intending to distinguish between patient and personal roles, which may be intertwined.

Problem-oriented medical/health record (POMR, POHR) or **problem-oriented record,** and **problem-orientation:** These concepts are explained in Chapter 7. See also Section 3.4.1 at Note 98.

Social determinants/drivers of health (SDoH): *SDoH* are factors such as housing, nutrition availability, financial status, social stresses, environmental conditions, access to health care, and more, that affect health. Recently, some have substituted the term "drivers" for "determinants," because social factors are not the sole determinant of health.

SOAP notes: The are progress notes in POHRs. *SOAP* is an acronym for Subjective, Objective, Assessment, and Plan, the four main components of a SOAP note. See Section 9.1.2.2.

PART I: THE PROBLEM

CHAPTER 2

Nature of the Problem

2.1 ROOT CAUSE OF THE PROBLEM

Neither the bare hand nor the understanding left to itself are of much use. It is by instruments and other aids that the work gets done, and these are needed as much by the understanding as by the hand. ... There is a single root cause of nearly all the evils in the sciences, namely, that while we wrongly admire and extol the powers of the human mind, we fail to look for true ways of helping it.

— Francis Bacon (1620)[13]

Patient care decision making depends heavily on the information taken into account. So decision making can be conceived in two stages: (1) assembling the right information in maximally usable form; and (2) making decisions after taking into account that information with the patient's preferences and values. The first stage is where the supply chain operates. Assembling information involves retrieval from the medical literature and other stores of knowledge, using that knowledge to select what patient data to collect and determine what the data mean for purposes of solving a medical problem as initially presented, organizing the results, delivering the results when needed, and recordkeeping to manage these processes and follow-up over time. All this must be individualized to the needs of each unique patient.

The primary vehicle for these supply chain activities has always been the doctor's mind. That role for doctors continues to a large extent even now that health IT is widely used. Yet, that role demands more than doctors' minds can deliver. This gap increases with advances in science and health IT. The advancing science increases the volume and complexity of medical knowledge exponentially. Corresponding increases in patient data arise. Health IT then worsens this overload on the mind by making access to information almost limitless and instantaneous. In short, the gap between the mind's limited capabilities and an effective informational supply chain has grown wider and deeper.

---

[13] Bacon, F., *Novum Organum* (1620), Summary of the Second Part, Aphorisms Concerning the Interpretation of Nature and the Kingdom of Man, Book I, Aphorisms No. 2, No. 9 (translated and edited by Urbach, P. and Gibson, J., Open Court Publishing Co, 1994). (Online versions of *Novum Organum* (in different translations from Latin) are available at various sites.) For a clinical psychologist's recent commentary on *Novum Organum* at the 400th anniversary of its publication, see Weinfurt, K., Francis Bacon's 400-year-old list of scientific foibles holds lessons for modern scientists. *Science* (March 17, 2020).

As this gap increased over time, it was never clearly recognized or planned for. "Most of the technologies and ways we do things in medicine were never designed with human limitations in mind," David Bates and Atul Gawande have written. "Indeed, most medical processes were never consciously designed at all; rather, they were built with a series of makeshift patches."[14] This point applies in spades to the informational supply chain. The outcome is ongoing failures of quality and economy on a pandemic scale. These failures have always been with us, so that we don't even recognize them as failures.

This state of affairs persists because the doctor's traditional role persists. Doctors are still seen as indispensable links in the informational supply chain.

The doctor's role is built into law (doctors have a legal monopoly over medical practice) and culture (we all are socialized to trust doctor expertise and authority). Moreover, the doctor's role distorts development and use of health IT. For example, doctors tend to use clinical decision support (CDS) tools selectively, when they feel the need for help with a puzzling case, not habitually, which is what scientific rigor requires. This tendency is the natural outcome of medical education and licensure, which implicitly teach pride in personal knowledge. Doctors are not trained to systematically recognize the inevitable discrepancies between accepted knowledge and problem solving for unique patients. Moreover, doctors are socialized to value professional autonomy. This makes them resistant to high standards of care for managing health information, which they see as bureaucratic interference in their autonomy.

Escaping this predicament means reconceiving the doctor's role and everything built around it, including standards for design and use of health IT.[15]

Think of the health care system as a network of nodes, human and organizational, playing their accustomed roles. This network is dysfunctional because its human nodes, doctors in particular, are severely underpowered and poorly connected.

The connectivity problem—poor interoperability among electronic systems—is well recognized. Less recognized is that doctors are underpowered for functioning as network nodes—their minds cannot reliably handle the information retrieval and processing functions required in ordinary medical practice. At the same time, doctors have a legal monopoly over medical decision making. Changing the division of labor is thereby blocked. Yet, all other nodes in the network continue to depend on doctors. That dependence threatens to become even more harmful to the extent that

---

[14]   Bates, D., Gawande A. Error in Medicine: What Have We Learned? *Ann. Int. Med.* Vol. 132, No. 9 (2 May 2000), pp. 763–767, at 764.

[15]   Although we thus advocate a large paradigm shift, we do not deny the potential for important advances within the current paradigm. See, for example, the impressive initiatives being disseminated by Dave Chase (as described at https://healthrosetta.org/ and his most recent book, *Relocalizing Health: The Future of Health Care is Local, Open and Independent*; the systemic accomplishments at the Geisinger health system as described in the book, Steele, G. and Feinberg, D., *ProvenCare: How to Deliver Value-Based Healthcare the Geisinger Way* (McGraw-Hill, 2019); and the management approaches described in the book, Emanuel, E., *Prescription for the Future* (PublicAffairs, 2017).

better interoperability increases data sharing. That would worsen the information overload that already burdens doctors intolerably.

With the right tools, so much of medical knowledge can become a network of interconnections, navigable by everyone as a map to the medical landscape. And so much of what doctors do with clinical information can be done better by clinicians and patients jointly equipped with the right tools. No machine learning or other AI is required; the tools simply perform transparent, pre-defined, step-by-step processes. Specifically, diagnostic and treatment decision making can be broken down into seven distinct steps, six of which can be automated in a completely transparent manner, using knowledge incorporated in the tools *before* patient encounters (see Section 8.3). No longer must medical practice be hidden inside a black box of "clinical judgment."

Doctors now recognize that they cannot rely solely on their own intellects for knowledge retrieval. But they still view themselves as key to the informational supply chain. Although tools external to the mind lessen the need for memorized knowledge, doctors believe that their minds are still required for *applying* that knowledge to unique patients. They believe their training and experience instill "clinical judgment"—a black box that no one outside the profession can open, a mysterious amalgam of science and art, based on acquired knowledge, intelligent analysis, and subtle intuition. And those judgments are sometimes impressive. The authority of doctors thus seems inherent in scientifically advanced medicine.

But this traditional role introduces constant breakdowns. Clinical judgment is a euphemistic term for doctors' on-the-fly, idiosyncratic, opaque cognitions, good or bad. Retrieving knowledge and matching it with patient data can occur reliably only if handled by tools external to the mind, *before* clinical judgment operates. Moreover, those tools avoid the waste inherent in requiring clinicians to reinvent the wheel for each patient, which is what happens when human minds perform information retrieval and processing that external tools could readily accomplish. That unnecessary burden on clinicians is not only wasteful but harmful, because it undermines their performance and their relationships with patients.

In short, habitual use of external tools should be an enforceable standard of care. There should be accountability for misplaced reliance on the minds of doctors. Human judgment has an essential role, but not as part of the informational supply chain—except in a limited, supplemental way, as discussed below.

Basic supply chain activities (knowledge retrieval and matching with patient data) are part of complex processes of care over time. Managing those processes is another part of the informational supply chain. It too requires external tools and standards of care for medical recordkeeping. Yet only fragments of the necessary standards have ever been recognized or implemented in paper or electronic records. The outcome is that current electronic record tools are in many ways dysfunctional (see Section 3.4).

Tools external to the mind offer not just transparency but productivity. Traditional reliance on the human mind in medicine is like a carpenter working without power tools. Power tools for carpentry offer not just speed and capacity but reliability and precision not possible with hand tools. Much the same can be said of digital information tools for handling medical knowledge and patient data. Moreover, digital tools for recordkeeping can be designed to structure and guide and coordinate the functioning of multiple users. They include not only doctors but other medical practitioners, consumers/patients/families, and third parties.

It follows that software developers need a healthy skepticism of the guidance they receive from doctors on the tools to be developed. Developers and users need to think in terms of new, higher standards of care than doctors are trained to deliver. Those higher standards must inform the design and use of the software tools to be built. The tools and standards are the subject of Part II of this book.

Skepticism of doctors is warranted because medicine's doctor-centric culture is anomalous. Unlike medicine, most fields in the domains of science and commerce have, over centuries, evolved away from reliance on opaque expert judgment.[16] The vulnerabilities of human judgment, and the importance of moving from subjective to objective knowledge, were recognized long ago in philosophy and psychology (see Chapter 6). Now, with digital information tools, the doctor's traditional role is a greater anomaly than ever. Until we face that reality, medical practice will continue to lack the rigor of scientific practice. And without that rigor, medicine's humanitarian ideals are hollow.

2.2   DENIAL OF THE PROBLEM

Four centuries ago Francis Bacon (see Note 12 and Section 6.1) recognized how human judgment naturally tends to misfire. Beginning a half century ago, that issue became the focus of research in cognitive psychology and then behavioral economics. In the late 1970s, LLW became aware of this psychology literature, discussing it in his writings, including a key 1981 article[17] and his 1991 book.[18]

Within the past three decades, doctors belatedly started paying attention to the psychology literature. Now clinicians regularly use terms from cognitive psychology such as heuristics, confirmation bias, recency bias, anchoring, premature closure, and the like—labels for the various ways that human judgment goes wrong.

---

[16]   See *Medicine in Denial* (Note 4), Part V, especially pp. 106–112 and 120–129.

[17]   Weed, L. L., Physicians of the Future, *New Eng. J. Med.* 1981;304:903–907 (see p. 905, discussing how "today's standard literature of psychology" demonstrates that "the typical physician is not and cannot ever be" capable of performing the tasks of medicine).

[18]   Weed, L., et al., *Knowledge Coupling: New Premises and New Tools for Medical Care and Education* (Springer-Verlag 1991), pp. 7–8, 350, citing, e.g., Tversky, A. and Kahneman, D., Judgment Under Uncertainty: Heuristics and biases in medical decision making. 1974. *Science*, 185:1124–1131, and Dawes, R., The Robust Beauty of Improper Linear Models in Decision Making. 1979. *Am. Psych.* 34(7):571–582.

But LLW diverged sharply from other doctors in the conclusions he drew. LLW concluded that cognitive psychology pointed in the same direction his work was headed. We cannot fully control how our innate tendencies compromise our judgment, but we can control how judgment is informed. The informed judgment of an ordinary mind may be superior to the uninformed judgment of a genius. So LLW found that we must *bypass* the human mind, replacing it with external tools, *for the first stage of decision making—assembly of information* (as explained in opening of Section 2.1). Our current reliance on the human mind for the first stage, where the informational supply chain operates, will come to seem as primitive as bloodletting.

In contrast, doctors conceive human judgment as if they can somehow learn to overcome its innate vulnerabilities. They are educated to rely on their own recall and reasoning powers to assemble the information they need. They are socialized to believe that "proficiency in clinical reasoning … is … the clinician's quintessential competency."[19] So now they have added cognitive psychology to their reasoning powers. Faith in the human mind thus remains deeply embedded in the culture of medicine.[20]

Dr. Sherwin Nuland articulated that faith, describing an ideal of intellectual virtuosity:

> To understand pathophysiology is to hold the key to diagnosis, without which there can be no cure. The quest of every doctor in approaching serious disease is to make the diagnosis and to design and carry out the specific cure. This quest I call The Riddle, and I capitalize it so there will be no mistaking its dominance over every other consideration. The satisfaction of solving The Riddle is its own reward, and the fuel that drives the clinical engines of medicine's most highly trained specialists. It is every doctor's measure of his own abilities; it is the most important ingredient in his professional self-image. … Our most rewarding moments of healing derive not from the works of our hearts but from those of our intellects—it is there that the passion is most intense.[21]

Relying on personal cognition to solve The Riddle is a false ideal. The falsity is that the ideal promises more than the mind can reliably deliver in real-world practice conditions. Not only does this ideal cause enormous medical and economic harm—it also contributes to toxic stress and burnout for clinicians. Beyond that, the ideal fosters harmful reward systems and blocks key regulatory and market reforms. Medicine's culture of denial is thereby perpetuated.

---

[19] National Academies of Sciences, Engineering, and Medicine. 2015. *Improving Diagnosis in Health Care.* Washington, DC: The National Academies Press, p. 53. https://doi.org/10.17226/21794. This same report observed that "clinical reasoning processes are difficult to assess because they occur in clinicians' minds and are not typically documented" in medical records (p. 94). It follows that "doctors' quintessential competency" is inherently resistant to scrutiny and continuous improvement by market and regulatory forces—a conclusion that the report did not articulate, which further evidences a state of denial.

[20] See Weed, L. L. and Weed, L., Diagnosing diagnostic failure. *Diagnosis.* 2014;1(1): 13–17. DOI: https://doi.org/10.1515/dx-2013-0020.

[21] Nuland, S., *How We Die: Reflections of Life's Final Chapter* (New York: Alfred A. Knopf, 1994), pp. 248–49.

A useful comparison is the culture of software development. Developers initially resisted the use of debugging tools, believing them to be merely a crutch that capable people should not need. But it rapidly became apparent that even the best developers needed these tools to cope with the complexities they faced. Use of the tools thus became standard practice in the software industry. In contrast, the medical profession has yet to establish comparable practices for medical decision making. This is true even with tech-savvy younger clinicians. Although they commonly use EHRs and clinical decision support software, they have never been subjected to the disciplined standards implemented in LLW's health record and CDS tools.

The persisting culture of denial is evident from "Lost in Thought," a 2017 *New England Journal of Medicine* (*NEJM*) article.[22] The authors lucidly describe how information overload has escalated in recent years.[23] Yet their article still displays denial of a simple reality—that medicine's complexity overtook the human mind long ago. The authors state, for example, "The complexity of medicine *now* exceeds the capacity of the human mind" (emphasis added). This is rather like saying, "The demands of transportation *now* exceed the capacity of horse-powered vehicles."

The authors similarly misstep in stating, "computers will *tomorrow* be able to process and synthesize data in ways we never could do ourselves" (emphasis added). Again, the reality is that computers have long had greater capacity than the human mind to "process and synthesize" data and knowledge in certain basic ways. This does *not* mean that computers can replicate the human mind—far from it. It means only that computers far exceed the human mind in reliability, speed, capacity, and precision when exhaustively performing determinate, definable tasks for purposes of information retrieval and processing. See Section 3.3.4.1.

The authors rightly point to recent advances in computer algorithms, machine learning, and data science, which enable new tools for pattern recognition and other indeterminate tasks. These tools are generating new clinical insights and approaches never before possible. But that focus neglects simpler capabilities that computers long since made possible. This neglect is part of medicine's culture of denial.[24]

A key reason for such denial, not specific to medicine, has been stated by the cognitive psychologist Robyn Dawes:

> The greatest obstacle to using [external aids] may be the difficulty of convincing ourselves that we should take precautions against ourselves … . Most of us … seek to maximize our flexibility of judgment (and power). The idea that a self-imposed

[22]   Obermeyer, Z. and Lee, T., Lost in Thought. *New Eng. J. Med.*, 2017; 377:1209–1211.

[23]   "Medical knowledge is expanding rapidly, with a widening array of therapies and diagnostics fueled by advances in immunology, genetics, and systems biology. Patients are older, with more coexisting illnesses and more medications. They see more specialists and undergo more diagnostic testing, which leads to exponential accumulation of electronic health record (EHR) data. Every patient is now a 'big data' challenge, with vast amounts of information on past trajectories and current states."

[24]   For further discussion, see *Medicine in Denial*, pp. 1, 25–27, 30–31, 115–120, 195–210.

external constraint on action can actually enhance our freedom by releasing us from predictable and undesirable internal constraints is not a popular one. ... The idea that such internal constraints can be cognitive, as well as emotional, is even less palatable.[25]

Another part of that culture of denial is rationalizing outright misdeeds, notwithstanding the serious harm they may cause. Misdeeds are committed by doctors and other elites in the health domain, just as corporate executives commit misdeeds in the business world. But these "familiar villains ... are not the fundamental problem," argues Dr. Robert Pearl. "They're what we call in medical practice "opportunistic infections," problems that turn up in the context of other diseases. Ridding the system of their misdeeds is not the ultimate solution. It won't significantly change the underlying pathology."[26] The underlying pathology involves imposing unbearable cognitive burdens on doctors while leaving them unconstrained by a system of accountability.

Dr. Pearl goes on to discuss classic research in psychology, showing

... that our environment has a far greater impact on our actions than our upbringing or personal beliefs. ... context—the circumstances we find ourselves in, the instructions we are given, the threats made against us, and the rewards we are offered—can and often do shift our perceptions of reality without our even recognizing that a shift has happened. Context has a profound impact on what we see, hear and feel. It has the power to change our behavior.[27]

Part of the context of medical practice is the absence of protective tools and standards external to the mind. This context is rooted in medical education and licensure for doctors.[28] The resulting environment fosters the "opportunistic infections" that Dr. Pearl speaks of—not just misdeeds by familiar villains, but constant missteps and shortfalls by everyone.

Denial is evident in an especially expensive arena for missteps—diagnostic testing procedures such as lab tests and imaging (CAT scans, MRIs, and others). Dr. George Lundberg has estimated that "about 80% of the tests carried out in the laboratories I oversaw in academic medical centers did not need to be done." He also cites studies of arbitrary doctor ordering behavior and observes: "doctors' examinations are now almost superseded by batteries of tests. When we look at why physicians order tests, we discover a wide variety of reasons, but few of them have anything to do

---

[25]   Dawes, R., *Rational Choice in an Uncertain World* (Harcourt Brace Jovanovich, Inc. 1988), p. 143.

[26]   Pearl, R., *Mistreated: Why We Think We're Getting Good Health Care—and Why We're Usually Wrong*. (PublicAffairs, 2017), p. 14.

[27]   Ibid. After discussing the psychology literature on terrible behaviors in non-medical contexts, Dr. Pearl goes on to provide various examples from medicine (pp. 15–19).

[28]   See *Medicine in Denial* (Note 4), part VIII.

with science."[29] A 2016 article estimates, "30% of laboratory tests and 20% to 50% of 'high tech' radiologic imaging ordered by health care providers may not be of value to the patient."[30]

As to imaging technologies, they are used promiscuously, with little forethought. Indiscriminate use is driven in part by vendor marketing and fee-for-service incentives, but also by analytical sloppiness and lack of transparency in "clinical judgment."

Beyond cost, the most obvious downside of imaging overuse is unnecessary radiation exposure for patients (depending on the technology involved). Other downsides involve clinicians not comprehending the fallibility of imaging results, not appreciating the enormous value of carefully selected, traditional data sources, and thereby becoming distanced from their patients and basic clinical data—not to mention losing (or never learning) important clinical skills.[31]

The very multiplicity of diagnostic approaches (imaging alternatives, traditional testing, basic clinical observations, and new genomics and other testing at the molecular level) make it difficult to know or recall the tests relevant to a given problem situation or to decide what tests to use when. Equally difficult is comprehending all the data generated. Yet, these difficulties could be greatly diminished by basic CDS tools of the kind described in Chapter 8.

Denial also takes the form of vested interests defending the status quo. A prominent example took place with an important publication about failures of quality. In late 1994, the *Journal of the American Medical Association* (*JAMA*) accepted for publication an article by Dr. Lucian Leape, "Error in Medicine" (now considered a classic[32]). But the *JAMA* editor, Dr. George Lundberg, was nervous. "I wanted to publish the paper for the profession but feared that I would lose my job if the public media hit hard on it." So Dr. Lundberg included the article in a *JAMA* issue published during the holiday season in late December, knowing that "holiday issues are the least read and covered." His tactic worked for a few weeks. Then "all hell broke loose. Hate mail began pouring in. I was accused of being on the side of the lawyers, of being a damned turncoat and traitor to the cause. An intensive lobbying campaign to get rid of me began" (but did not succeed at the time).[33]

Meanwhile, 18 days before the date of the *JAMA* article, unbeknownst to Dr. Leape and Dr. Lundberg, a shocking medical error had taken place:

> Betsy Lehman was a nationally recognized *Boston Globe* health columnist and mother
> of two young girls when she died of a massive overdose of chemotherapy while being

---

[29] Lundberg, G., *Severed Trust: Why American Medicine Hasn't Been Fixed*. (New York: Basic Books, 2000). In one of the studies cited, the volume of orders of two expensive tests was reduced by two thirds and one third merely by changing the hospital lab request form. ("Just as most diners rarely order something that is not on the menu, so doctors rarely order tests not listed on the test request slip.") Ibid. pp. 22, 257–59. So much for sophisticated clinical judgment.

[30] Litkowski, P. et al., Curbing the Urge to Image. *Amer. J. Med.* 2016;129(10):1131–1135, DOI: https://doi.org/10.1016/j.amjmed.2016.06.020.

[31] For more detailed discussion, see *Medicine in Denial*, Note 4, pp. 86–87, 125.

[32] Leape, L., Error in Medicine. *JAMA* 272;23:1851–1857 (Dec. 21, 1994).

[33] Lundberg, G., *Severed Trust: Why American Medicine Hasn't Been Fixed* (Basic Books, 2000), p. 171.

treated for breast cancer at the Dana-Farber Cancer Institute on December 3, 1994. … Betsy Lehman's death catalyzed a movement to recognize that patient harm is not always caused by an individual clinician's negligence. Rather, *preventable medical harm can be viewed as a consequence of institutional systems and culture that had not kept pace with the complexities of modern health care.* The challenge and the opportunity, then, would be to apply interventions developed by other complex, high-risk industries that had succeeded in achieving high levels of safety and reliability.[34]

Subsequent reporting of this incident combined with Dr. Leape's *JAMA* article began to break through medicine's culture of denial. Health care providers and the media increasingly recognized threats to patient safety and the need for "institutional systems and culture" changes to better cope with medicine's escalating complexity. This was soon followed by increasing recognition of the human mind's vulnerabilities as shown by cognitive psychology. But this recognition went only so far. It did not extend to re-thinking medicine's occupational hierarchy with the minds of doctors at the top. On that front, denial goes deep. See Chapter 10.

A doctor-centric culture is naturally in denial that the doctor's role is itself a problem. And every year, the culture receives an infusion of newly minted doctors who have been indoctrinated and licensed to perform the doctor's role while being insulated from real competition. The effect is to block demand for solutions that might otherwise be rapidly embraced. The culture is thus self-reinforcing.

The culture of denial extends to patients. "Some people don't want to hear about medical mistakes or don't believe they can happen. One person says, 'People put a lot of wishful thinking into their doctor and the system, and they don't want to hear about what can go wrong.'" To hear what can go wrong is to hear "stories of people who have lost a part of themselves. That part may be a loved one, or a physical part of their body, or their identity and sense of self."[35] Hearing such stories hardly generates demand to hear more.

Denial also occurs in the health IT world, where software developers are expected to satisfy doctors or business executives or both. This environment has led to a health policy consensus that does not threaten the doctor-centric status quo. The consensus is that health IT specialists need to fix poor "interoperability" among different computer systems. This is indeed a major concern— without sufficient interoperability, electronic documents and data transmitted from one system to another are not "computable" by the receiving system and thus less easily usable by clinicians and

---

[34] Betsy Lehman Center for Patient Safety: The Financial and Human Cost of Medical Error … and How Massachusetts Can Lead the Way on Patient Safety (June 2019), p. 5 (emphasis added). The medical error in Betsy Lehman's case (a chemo overdose) was not an error of decision making (the subject of this book) but an error of execution (largely outside the scope of this book). See Section 1.1 and Note 3.

[35] Gibson, R. and Singh, J. P., *Wall of Silence: The Untold Story of the Medical Mistakes That Kill and Injure Millions of Americans* (Lifeline Press, 2003), p. 16.

patients.[36] But many seem to hope that interoperability alone will somehow enable new, self-organizing solutions to medicine's manifold problems. Such hopes for interoperability reflect denial of a broader and deeper problem—defective information processing by the human minds at both ends of interoperable information exchange, those who generate the information and those who request and receive it.

An example is doctors' thoughtless ordering of diagnostic testing procedures such as lab tests and costly imaging, discussed above at Notes 29 and 28. Interoperability does help reduce duplication of these procedures by different providers, but that's not enough to assure that ordering of these procedures or use of the results is rational.

As a further example, suppose a doctor uses interoperable medical record systems to request transmission of the problem list from some former doctor's EHR. Interoperability might assure that the receiving record system properly adds this data transmission to the problem list in the receiving doctor's record for the right patient. But the transmitted list must be reconciled with the problem list already maintained in the receiving record—an issue that interoperability does not address. Nor does it address a more fundamental issue: no one has assurance that either doctor's problem list is complete or the problems carefully defined. The underlying failure is that standards of care for problem lists have yet to be generally accepted, much less enforced. Medicine persists in denial of the reality that enforcing those standards with corresponding tools is needed to "take precautions against ourselves."[37]

We recognize that unaided clinical judgment is sometimes remarkable, in terms of both analytic insight (thoughtfully applying first principles to novel situations) and intuition (pattern recognition at the unconscious level). This typically happens when well-trained, highly experienced, meticulous, intelligent clinicians bring their minds to bear on difficult cases, especially those that happen to match well with the clinician's personal expertise. But clinicians of that caliber are costly. There will never be enough of them. Moreover, they tend to be so overburdened that they lack time to meticulously apply their expertise. In any event, they are still fallible, and matching their expertise with each patient's needs is essentially random. Capable doctors are no substitute for a true system of care.

No system can capture all the human intangibles involved in caring for patients. But a well-conceived system would capture and disseminate and continuously improve much of what the best doctors—*and other clinicians*—collectively have to offer. Recall the wise physical therapist whose special knowledge rescued Dr. Topol from the nightmare with his orthopedist.

---

[36]   What should be accepted as sufficient interoperability is a matter of debate. For a valuable recent discussion of the excruciating complexities involved, see Adler-Milstein, J. et al., Improving Interoperability By Moving From Perfection To Pragmatism, *Health Affairs* Blog, Jan. 13, 2021. DOI: 10.1377/hblog20210105.661344.

[37]   Dawes, Note 25.

CHAPTER 3

Examples of the Problem

Decision making in medicine is too often a hit-or-miss process, full of unnecessary trial-and-error. Below are examples of the countless breakdowns involved. These examples illustrate why medical decision making cries out for tools and standards external to the doctor's mind.

Some readers may object that these and other examples in this book are cherry-picked to illustrate breakdowns. To those readers, a more representative selection of examples would show that medicine succeeds more often than not, with health professionals struggling heroically to achieve those successes. They are right. But those realities of success co-exist with the realities of failure. The examples below are not just anecdotes. They are "sentinel events," signaling failures of enormous magnitude (see Chapter 4). They are clues to the realities that quality standards in medicine are lax, that much of the trial-and-error and suffering in medicine are avoidable, that the struggle and burn-out endured by health professionals are intolerable, that the waste of human and financial resources is unacceptable, and that reclaiming those resources for more productive use is essential if we are ever to overcome other crises the world faces.[38]

3.1 DIAGNOSTIC ODYSSEYS

An all too frequent occurrence in medicine is avoidable "diagnostic odysseys." These are cases where the correct diagnosis is unnecessarily delayed or missed altogether, as the patient undergoes countless procedures, some of them invasive and risky, and incurs great costs, economic and psychic. In these cases multiple clinicians fail to recognize a known disease, even though the disease would be identifiable with use of simple information tools. (These avoidable cases are to be distinguished from unavoidable cases where the correct diagnosis is either unknown to medical science or inherently difficult to recognize even with the best tools and processes.)

That these avoidable failures still occur as much as they do is astonishing. Avoidance simply requires matching up data with knowledge—an initial data point (the presenting problem) with knowledge about what data to collect for investigating that problem, and then knowledge about what the collected data mean. This matching is readily accomplished with basic computer software tools, as distinguished from advanced AI.

These avoidable failures arise from a dysfunctional division of labor between the human mind and external information tools. The failures occur in treatment as well as diagnostic decision making (see Section 3.2). Both types of decision need to be transformed from judgment-driven to tool-

---

[38] See Section 4.2 on the economics of quality failures.

driven processes. The necessary tools should guide the human mind in two basic functions: applying medical knowledge to patient data, and using structured health records to organize and communicate the processes of care over time. The tools are described in Chapters 7–9.

This section provides a few examples involving diagnosis. In considering these examples and the processes involved, keep in mind the following conclusion from a study of diagnostic error:

> Most errors were related to patient-practitioner *clinical encounter-related processes, such as taking medical histories, performing physical examinations, and ordering tests.* … preventive interventions must focus on common contributory factors, particularly those that influence *the effectiveness of data gathering and synthesis in the patient-practitioner encounter* [emphasis added].[39]

3.1.1   TWELVE YEARS OF MISERY

Pelvic pain in women has a number of possible causes. One is endometriosis, a disease where tissue normally lining the inside uterine wall grows outside the uterus. This condition is common and well known as a cause of pelvic pain. Yet it often goes undiagnosed. The following summarizes such a case.[40]

A healthy 15-year-old girl developed disabling menstrual cramps and heavy bleeding with her first menstrual period. A gynecologist immediately started her on an oral contraceptive pill as a treatment for "bad periods."

For the next three years her menstrual cramps continued, and she developed severe abdominal cramps, bloating, nausea, and diarrhea. Her gynecologist continued her on oral contraceptives. She saw her primary care doctor for her severe diarrhea; his diagnosis was Irritable Bowel Syndrome with Diarrhea (IBS-D). This became a *de facto* final diagnosis (as distinguished from a preliminary "working" diagnosis), because no additional diagnostic testing was pursued at the time. She obtained a second opinion from another gynecologist who also attributed her symptoms to "bad periods," again without additional testing.

Nine years after her initial menstrual pain began and six years after her disabling abdominal pain and diarrhea began, this girl was referred to a gastroenterology (GI) specialist. The specialist performed a colonoscopy that showed only a "tortuous bowel." The GI doctor told her this finding confirmed the IBS-D diagnosis.

---

[39]   Singh, H., Giardina, T. D., Meyer, A. N., Forjuoh, S. N., Reis, M. D., and Thomas, E. J., Types and origins of diagnostic errors in primary care settings. *JAMA Intern Med* 2013;173:418–24. DOI: 10.1001/jamainternmed.2013.2777.

[40]   The case summary is paraphrased from Mackenzie, M. and Royce, C. "Endometriosis: A Common and Commonly Missed and Delayed Diagnosis," Agency for Healthcare Research and Quality/Patient Safety Network, published June 2020. The case summary is followed by commentary, quoted below.

Three years later, the girl experienced "sharp" right-sided abdominal pain. At the emergency room a computed tomography scan was interpreted as showing acute appendicitis. She then had an emergency appendectomy by a general surgeon. At her postoperative appointment, the surgeon informed her that endometriosis lesions close to the appendix, which prolonged the surgery, had caused the appendix to become infected. The surgeon did not call in a gynecological surgeon during the surgery or complete a biopsy. Nor did he remove any of the suspected endometriosis as he feared causing further spread of endometriosis cells. The surgeon referred her to a gynecologist.

The girl experienced often intolerable medication side effects from hormonal treatments—metabolic changes with weight gain as well as psychological and emotional derangements including anxiety and depression. Ultimately, diagnostic laparoscopy confirmed endometriosis via biopsy, and definitive laparoscopic surgery removed the endometriosis tissues. This happened 12 years after the girl's symptoms started.

Looking back, the girl felt that from the beginning that her symptoms were often dismissed as psychological. She was made to feel that "she was crazy." Without a strong advocate (her mother), she felt that doctors would not have pursued additional diagnostic testing and effective treatment.

The above summary of the case was followed by detailed commentary, beginning with "Missed Opportunities for a Timely Diagnosis and Treatment." Some of the missed opportunities arose from disorganized decision making and recordkeeping. Specifically (citations omitted):

- "It is impossible to know if the first gynecologist considered endometriosis; since the initial treatment for endometriosis consists of non-steroidal anti-inflammatory agents and combined oral contraceptives, it is possible that these medications were prescribed for symptom relief and as empiric treatment of possible endometriosis. Premature closure on the diagnosis of dysmenorrhea resulted in no further investigation…. "

- "The second gynecologist also contributed to the delayed diagnosis. This consultant clearly accepted the working diagnosis of primary dysmenorrhea and did not pursue further diagnostic testing. The cognitive bias of anchoring (wherein first diagnostic impressions persist or are even cemented despite contradictory evidence) seems to have prevented consideration of any other explanation [citation omitted]. To be fair, however, the same hormonal management or variations of it—progesterone-only pills, implants, shots or intrauterine devices—would have been a standard treatment for either primary dysmenorrhea or endometriosis."

- After IBS-D diagnosis, "no further workup was performed for several years," which meant that "other infectious or inflammatory gastrointestinal conditions could have been missed."

- The GI doctor later performed a colonoscopy and noted a "tortuous bowel." He misinterpreted this finding as confirming the IBS-D diagnosis. In fact, "tortuosity is a nonspecific finding of uncertain significance, and its link to a gynecologic condition … probably wasn't recognized by the gastroenterologist as a cause of GI symptoms."

Further commentary included the following points (citations omitted):

- "Average delays [in diagnosis of endometriosis] range from 6 to 11 years, often despite disabling and ongoing symptoms, with symptoms being 'treated' for years despite patients not having received a definitive diagnosis [citations omitted]. Endometriosis poses a particular diagnostic challenge … Nonetheless, the most commonly recognized etiology for chronic pelvic pain is endometriosis, adenomyosis and associated spasms of pelvic floor muscle."

- "Given this possibility of multi-organ involvement, patients with endometriosis are often referred to multiple specialists, contributing to a delay in diagnosis while alternative etiologies are considered. Patients sometimes describe taking a 'grand tour' of specialists including not just gastroenterologists but endocrinologists for metabolic syndrome/polycystic ovarian syndrome, rheumatologists for undifferentiated inflammatory conditions, neurologists for neuropathic conditions, and orthopedics for chronic back pain, leg pain or even shoulder pain. After many years of unremitting pain with no apparent cause (and sometimes earlier in their diagnostic journey), patients are commonly referred to psychiatrists for anxiety and depression, often as a consequence rather than the cause of pain. …"

- "According to the patient in this case, perhaps the most damaging aspect of the disease was not the symptoms or the side effects of treatment but rather the persistent dismissal of her symptoms as 'normal.' This experience of 'victim-blaming'—the implication that her symptoms were psychogenic or in some way self-induced—is commonly reported by endometriosis patients. Viewed in this light, the associated anxiety and depression are not unexpected."

- "Alienation from the healthcare system due to emotional distress is common and may further delay diagnosis. Many endometriosis patients have an intuitive sense that something is desperately wrong as they suffer from persistent pain and dysfunction. Repeated dismissal of their symptoms and their experience is understandably devastating."

- "On a health systems level, the historic and systemic under-investment in women's health - from education of trainees to research in effective treatments for conditions unique to women - magnifies the alienation and distrust that can develop from professional disregard of the lived experience of women who feel the pain of endometriosis and have often long suffered with its symptoms. The evaluation of adolescents with severe dysmenorrhea—who may have difficulty advocating for themselves—represents another realm in which healthcare professionals need to listen to and believe their patients."

The commentary concluded with a number of "take-home points," which included the following:

- "The prolonged average time interval from symptom onset to diagnosis of endometriosis is the result of fundamental misconceptions and gaps in knowledge regarding the disease, its pathogenesis, natural history, presentation, limited utility of imaging, and treatment."

- "Because endometriosis often involves multiple organ systems including GI, genitourinary, peripheral nervous, and respiratory systems, patients are often referred to specialists who need to consider this diagnosis. Likewise, primary care doctors and gynecologists need to fully appreciate the myriad 'non-gynecologic' manifestations of this disease."

- "Accurate diagnosis of endometriosis requires active and empathetic listening to patients' complaints, informed by an understanding of the presentation and additional medical history of the patient and, most importantly, avoidance of "normalizing" pain."

3.1.2   TUNNEL VISION BY TEN DOCTORS IN FIVE SPECIALTIES[41]

For three years Carol Hardy-Fanta consulted a series of doctors about her medical problem. The problem was repeated falls for no apparent reason. None of the ten doctors—her internist, four orthopedists, three neurologists, a rheumatologist, and a podiatrist—could arrive at a successful diagnosis. They did come up with various diagnostic hypotheses, but none of those explained all her symptoms or led to successful treatment. Ms. Hardy-Fanta herself researched her problem, identified the correct diagnosis as a possibility, and pushed her doctors to consider her idea after they initially rejected it.

At the time her falling problem began, she had been experiencing hip pain, foot pain, and a change in her stance. The podiatrist prescribed a walking boot, which seemed to worsen her hip pain. Her internist diagnosed bursitis, which would account for the hip pain and possibly the change in her stance. He recommended physical therapy. But her pain continued, so she next consulted a rheumatologist. He ordered MRI scans and found only mild hip arthritis. Then she went

---

[41]   This case is reported in an April 20, 2020 article, *She fell more than 30 times. For three years, doctors couldn't explain why,* from the *Washington Post* Medical Mysteries series.

to an orthopedist, who prescribed cortisone shots. A series of doctors gave her five shots in the next seven months, more than the maximum safe amount.

None of this significantly relieved her pain. Meanwhile, her falling problem continued, and new symptoms appeared. Her left hand was sometimes involuntarily clenched, her handwriting became smaller, at times illegible, and her speech became soft, sometimes to the point that her husband had trouble understanding her. Then she consulted a neurologist, who found "no evidence of a neurologic cause" of her symptoms. He recommended gait training for the falling problem and continued physical therapy for the hip pain.

During this process, Ms. Hardy-Fanta (herself a distinguished academic) conducted her own research, which led her to suspect Parkinson's disease. She consulted a second neurologist, but he also saw "no evidence to suggest an underlying neurological disease." He dismissed Parkinson's, based on absence of characteristic symptoms he expected (tremor, rigidity, slowed movements).

Meanwhile, the hip pain worsened, and the falling problem continued to the point where Ms. Hardy-Fanta sometimes used a walker or wheelchair to prevent falls (in one fall, she broke her arm, which required surgery). Then an MRI showed dead bone tissue in her right hip. This required a total hip replacement—a major procedure that did provide temporary pain relief.

But the problem with her clenched left hand got worse, and she developed a gait problem commonly seen with Parkinson's disease. She went to another neurologist, who diagnosed a "frontal gait disorder" and ordered a PET scan to rule out a rare type of dementia. The scan was negative. Then that neurologist "suggested that if Hardy-Fanta was concerned about Parkinson's," she could try a specialized brain scan to check for reduced dopamine activity, which is associated with Parkinson's. She had the scan, and its results were indeed "consistent with" Parkinson's. She went to a third neurologist, who prescribed a drug commonly used for Parkinson's symptoms. Her condition improved, further supporting Parkinson's as a diagnosis (taking the drug without resulting improvement would suggest a different movement disorder than Parkinson's).

The *Washington Post* article reports that Ms. Hardy-Fanta was

> angry about what she regards as tunnel vision: The orthopedists seemed to attribute her pain to an orthopedic cause when it was likely a sign of Parkinson's. And the neurologists had ruled out Parkinson's because she did not display several hallmarks of the disease, including a tremor or rigidity, even though her soft speech, handwriting changes and clenched hand are common signs of the disease.

Moreover, she suspected that this delayed diagnosis might have led to the hip replacement (her hip bone damage might have resulted from the excess cortisone shots, which would not have been prescribed had Parkinson's been recognized). "It is common … for Parkinson's patients to undergo unnecessary orthopedic procedures, including surgery, for symptoms that turn out to be related to the neurological disease and not a musculoskeletal problem," according to the *Post* report, citing Dr.

Michael Okun, a neurologist who directs the Parkinson's Foundation. "We need to do a better job" making the diagnosis, Okun acknowledged to the *Post*.

Dr. Okun also commented on lack of knowledge of how the disease affects women. Several reader comments on the *Post* article observed that negative stereotypes about women by male doctors contribute to misdiagnosis of Parkinson's (e.g., "Women are more likely to be blown off as 'psychosomatic' than men"). Such stereotypes fill the vacuum left by sloppy diagnostic practices. The outcome is "alienation and distrust that can develop from professional disregard of the lived experience of women," as described in the endometriosis case above.

This case illustrates several further characteristics of these diagnostic odysseys.

- Clinicians begin by considering only the diagnostic possibilities that come to mind. Each clinician's initial list of possibilities (known as the "differential diagnosis") is different, and no one's list is complete.

- Specialists frequently have "tunnel vision," either because they fail to realize how often actual patient problems cross specialty boundaries, or because they take no responsibility for anything outside their specialties. The result is that what happens to a patient depends on what specialists are consulted and when. And there is no assurance of consistency even within a given specialty. For example, after Ms. Hardy-Fanta's first two neurologists found no evidence of neurologic disease, her third neurologist suggested a rare type of dementia as the diagnosis. After a brain scan did not support that hypothesis, he conceded that she could try a test for her Parkinson's hypothesis, which turned out positive. Then she went to yet another neurologist for treatment.

- A common error is ruling out a disease because its hallmarks (expected symptoms and signs) are missing. Doctors fail to realize the extent to which a given disease can manifest itself quite differently from one person to another over time, with the result that many patients will not have some of the expected hallmarks of the disease at any one time (or ever). Two basic reasons for this variability are the following.

  - Every person is unique, causing a disease to interact variably with people who differ in their physiology, psyche, and circumstances.

  - A single disease label may be applied to variable cases or else to seemingly similar cases that are in fact different diseases with different causes and pathologies. See, for example, this explanation of Types of Parkinsonisms.

- Diagnostic misconceptions often lead to misguided treatments. Doctors are too quick to treat unproven diagnostic hypotheses as a basis for treatment decisions, even when the chosen treatments are risky, as happened with Ms. Hardy-Fanta's cortisone shots.

In effect, treatment attempts become diagnostic tests; if a treatment is followed by improvement, then this result is thought to confirm that the patient has the disease for which the treatment is normally given. Sometimes referred to as "empiric therapy," this accepted practice may be the best available approach, as ultimately happened for Ms. Hardy-Fanta (when she experienced some improvement after receiving a treatment for Parkinson's). But such an approach to diagnosis is not trustworthy unless all other alternatives are considered. Absent that thoroughness, empiric therapy creates risk that its apparent success is coincidental or temporary; better alternatives may have been overlooked.

The above points illustrate some of the ways the human mind can go wrong. They also illustrate the lack of established standards and tools to cope with that fallibility in an organized way. The result is that care is driven by clinician idiosyncrasies instead of patient needs.

3.1.3 RANDOM ROADS TO THE SAME DIAGNOSIS

3.1.3.1 Missing a Girl's "Obvious" Diagnosis Until She "Almost Died"[42]

A 15-year-old girl was admitted to a teaching hospital with "excessive fatigue." This problem had made her unable to attend school for the prior three months. For the seven months before admission, she had experienced absence of periods, weight loss, and shortness of breath. Her initial exam at the hospital showed mildly low blood pressure. She went home and then was re-admitted to the hospital three times in the next month with additional symptoms. These included upper abdominal pain, other diffuse abdominal pain, nausea, bile vomiting, diarrhea, dehydration, further weight loss, and a rapid heartbeat episode. Numerous skin moles were also commented on by several practitioners but were apparently not viewed as relevant. Other findings were normal except for some borderline blood test results.

As often happens in such diagnostic odysseys, psychiatric causes were considered. An endocrinologist suggested an eating disorder. A urine test suggested use of ipecac and thus bulimia (bulimics sometimes use ipecac to induce vomiting) or poisoning by a family member. Despite adamant denials by the girl and her family, referral to child protective services was considered, and

---

[42]   This case is reported in Keljo, D. and Squires, R., Clinical Problem Solving: Just in Time. *N. Engl. J. Med.* 1996. 334:46-48. Multiple letters to the editor on this article appear at *N. Engl. J. Med.* 1996. 334:1403–1405 (May 23, 1996). A detailed analysis of the article and the letters appear in our book *Medicine in Denial* (Part II.A, pp. 15–28, with further analysis at pp. 29, 88–89, 100, 102–103, 180, 185, which the above discussion draws on. For an example of a similar case with a different outcome, see UK Girl Dies After Doctors Fail to Diagnose Addison's Disease (July 29, 2014). The coroner's inquest in that case concluded, "Clare exhibited significant symptoms and red flags to warrant further investigation and, in all likelihood, the investigation would have led to her diagnosis." See also the next section discussing a very recent case of Addison's disease with abnormal pigmentation.

a psychiatric evaluation was ordered. But the clinicians abandoned their psychiatric theories. Follow-up urine tests suggested the original test result was erroneous), and the psychiatric evaluation did not support the eating disorder theory.

Finally, months after the first hospital admission, further blood testing showed a below-normal serum sodium level (which initially had been one of the borderline blood test results) and other blood electrolyte abnormalities. This suggested Addison's disease (a lack of adrenal-cortical hormones that is fatal if left untreated).[43] Follow-up tests confirmed this diagnosis, and the disease was successfully treated by putting the girl on hormone replacement therapy.

But that outcome did not occur "until it was nearly too late to save the child's life." In the meantime, she endured "vastly excessive testing," including "dozens of blood tests, immunologic studies, endoscopies, other radiographic tests and biopsies," plus futile attempts at treatment. "Fortunately, the patient survived not only her illness but the myriad tests and treatments administered," the article observes, acknowledging the debilitating effects of all that "care." The article does not state how long this process took, but it appears to have lasted for several months after the initial hospital admission (which followed seven months of symptoms). The article also makes no mention of the psychiatric symptoms that commonly appear with Addison's disease. Nor does the article describe the enormous stress that the girl and her family must have endured. Nor does the article quantify all the clinician time invested in the case.

Yet, most of this futility was avoidable. The correct diagnosis was "obvious in retrospect," given telltale signs present at or near the outset of care. "Fatigue, weakness, dehydration and hypotension [low blood pressure] are classic manifestations of Addison's disease. … Multiple practitioners commented on the patient's large number of deeply pigmented nevi [skin moles], and there is a report of such changes in Addison's disease," the authors belatedly admitted, citing a 1990 article on a case in Denmark.[44] Yet these "multiple practitioners," after observing this abnormal pigmentation, missed its significance. Also indicative of the disease were several of the findings made in the first month after the initial admission. Yet none of the specialists involved recognized the pattern. Moreover, the below-normal serum sodium that finally led to the diagnosis could have been recognized during the first month of care. At that point the level was borderline but should have been viewed as below normal because the girl was dehydrated (dehydration commonly affects blood tests, including serum sodium).

---

[43]   Addison's disease is now known as primary adrenal insufficiency (PAI). We use the Addison's disease label, since that term is used in the *NEJM* article

[44]   Authors' reply to letter to the editors (see Note 41). This girl's abnormal pigmentation was not an unusual symptom of her disease. "*The most specific sign* of primary adrenal insufficiency is hyperpigmentation of the skin and mucosal surfaces, …" (emphasis added). Betterle, C. et al., Autoimmune adrenal insufficiency and autoimmune polyendocrine syndromes: Autoantibodies, autoantigens, and their applicability in diagnosis and disease prediction. 2002 *End. Rev.* 23(3):327–364, at p. 331.

Such chaos has a predictable outcome: "vastly excessive testing and numerous attempts to treat putative diagnoses are the rule. *We can be certain that in such instances some patients die because the correct diagnosis is never entertained and that even after an autopsy the mystery often persists*" (emphasis added).

This case cannot be viewed as an outlier reflecting unusually deficient practice. On the contrary, the case involved multiple specialists at an academic teaching hospital, where one would expect unusually skilled practice. The article was published in one of the world's leading medical journals, where it was presented as a "perplexing" case. "Only the toughest critic could fault any of the clinicians for not making the correct diagnosis earlier," the authors concluded. The letters to the editor expressed differing views on fault, with no consensus on how the case should have been handled. In response to criticism that the lists of diagnostic possibilities considered were incomplete, the authors simply responded, "clinicians work from short lists, and these lists vary from specialty to specialty."[45]

A clue to why the case seemed so difficult lies in how the doctors assembled the information leading to the diagnosis. The article appeared in the *NEJM*'s "Clinical Problem-Solving" series. In these articles, patient data is presented "in stages" to an expert clinician, who analyzes the data presented at each stage.[46] So the article gradually assembled and presented the data piecemeal, mimicking what doctors actually do in practice. A completely different approach would be first to collect detailed data all at once and up front, with the needed data being defined in advance for the presenting problem of fatigue. Then all the data could be analyzed *in combination*. The difference between these two approaches is fundamental. A tool designed for the second approach makes it entirely feasible in real-world practice.

Such a tool requires *habitually* using the tool up front to collect and analyze all data indicated by the tool *before jumping to any conclusions*. This high standard of disciplined behavior, not personal knowledge, is what the scientific rigor demands. That disciplined behavior should be treated as an enforceable standard of care. It follows that the doctors who "cared" for this girl should be faulted not for ignorance, which is inevitable, but for trusting their personal knowledge.

Had that standard been followed here, then the case would have seemed easy, not perplexing. Indeed, had the 15-year-old patient and her family themselves been equipped with the right tool, they might themselves have used it to enter the symptomatic findings, gone to a clinician to enter the physical exam and lab findings, and then easily identified Addison's disease as a likely diagnosis.

---

[45] *N. Engl. J. Med.* 1996, 334: 1403–1405 (authors' response to letters to the editor about the Addison's disease case discussed in Section 3.1.3).

[46] The articles in this series are prefaced by the following: "*In this Journal feature, information about a real patient is presented in stages (boldface type) to an expert clinician, who responds to the information by sharing relevant background and reasoning with the reader (regular type). The authors' commentary follows.*" Thus, a leading medical journal perpetuates a defective method for the initial workup—proceeding "in stages."

In short, whether such a case is easy or perplexing depends on the tools and standards used or not used for assembling the relevant information.

3.1.3.2 "These Days She was Queasy All the Time. … She Always Felt Tired."

Another case of Addison's disease illustrates the power of using an external tool but also the limitations of a tool without standards of care for its design and use. Consider an October 2020 article by Dr. Lisa Sanders in her *New York Times Magazine* series on diagnosis, She Craved Salt and Felt Nauseated for Months. What Was Wrong?. In that case, a 46-year-old woman suffered a variety of symptoms, such as fatigue, salt craving, weight loss, and gastrointestinal symptoms (the article begins by describing a traumatic episode of severe vomiting in a restaurant). Her primary care doctor was useless in making a diagnosis (he suggested acid reflux).

Then she noticed dark spots on her face and hands. So she went to a dermatologist in the hope of at least getting help with this skin condition, if nothing else. The dermatologist asked her about sun exposure and tanning habits. Then, "he stepped out of the exam room and went to his office. At his computer, he searched for causes of hyperpigmentation. Two rare disorders came up immediately"—hemochromatosis and Addison's disease, neither of which the dermatologist had ever encountered. He ordered lab tests, telling his patient that Addison's disease was "a long shot." But the test confirmed Addison's, and the dermatologist referred the patient to an endocrinologist, who promptly initiated hormone replacement therapy. She started recovering within 48 hours.

That hyperpigmentation is a common symptom of Addison's disease has been known for decades (although it was not known to the doctors in the other Addison's case discussed above). That diagnostic link was uncovered by the dermatologist not because of his scientific knowledge but because of his scientific behavior—using an external tool to go beyond his personal knowledge. This doctor thus did what scientists habitually do but doctors are not educated or required to do by their training and licensure. Medical students "are stuffed with the facts of basic science, but the behavior of a scientist escapes them …"[47]

This doctor's success with an Internet search does not mean that tool is sufficient for diagnosis. For example, occasionally hyperpigmentation does not appear with Addison's disease.[48] Or (more likely) it might not have yet appeared at the time of a patient-clinician encounter (which may be what happened in this patient's encounters with her primary care doctor). So the diagnostic process should not focus on any one symptom at any one time. What is needed is a tool optimized for handling variable symptom *combinations* and doing so *over time*. See Section 3.3.2 and Chapter 8.

---

[47] Weed, L. L., *Physicians of the Future* (Note 16), p. 906.

[48] There are "a few reports of Addison's without hyperpigmentation," according to an article citing a dozen references on the point, Teran, S. et al., An atypical case of familial glucocorticoid deficiency without pigmentation caused by coexistent homozygous mutations in MC2R (T152K) and MC1R (R160W), *J. Clin. End. Met.*, Vol. 97, Iss. 5, 1 May 2012, pp. E771–E774, DOI: https://doi.org/10.1210/jc.2011-2414.

3.1.4   "I DREAD EVERY ANNUAL MAMMOGRAM"—OVERDIAGNOSIS AND RELATED HARMS

The above cases involve missed and delayed diagnoses, where the informational supply chain is ineffective in transmitting key information via the minds of doctors. That supply chain can also become clogged with diagnostic activity that is useless, costly, and sometimes harmful. A prime example is screening large populations for unidentified disease. Dr. H. Gilbert Welch describes such a case involving mammography screening for breast cancer. "When I have a piece pointing out the limitations of mammography in the general press," he writes, "I get letters like this":

> "I am a 66-year-old woman who has had a difficult experience with mammography over the past 20 or so years. For some reason, I have a strong tendency to develop calcifications, most of which they feel the necessity to biopsy. I dread every annual mammogram because the likelihood is very high that something will have to be checked out. So far nothing has been wrong, but I have had one open biopsy and three stereotactic biopsies. The last of those biopsies produced an incidental finding of a papilloma, which they decided to do a "lumpectomy" on because one in ten can hide cancer. The surgery did not go well; they informed me they missed the spot and would have to redo the surgery. Two days later, they decided they had indeed operated on the right area and that there was no cancer in the papilloma. No cancer, but extreme trauma to the patient and a developing panic problem with regard to the whole issue for which I will now be seeing a psychologist." [49]

Dr. Welch presents this case in arguing that screening for disease tends to cause three types of harm: fear, false alarms, and overdiagnosis. Fear occurs when the screening itself, or false alarms, creates fear of the disease being screened for. The resulting stress constitutes an adverse side effect of screening. As to the harm of overdiagnosis, this does not mean excessive diagnostic activity. Instead, overdiagnosis occurs when screening uncovers a disease that will never cause harmful symptoms or death and thus never require treatment. The difficulty is that often no one can know in advance whether the disease should be treated, or monitored, or ignored (the patient may be left with a burden of fear regardless). Such uncertainty often occurs with prostate cancer diagnosis, for example.

Unlike treatment, which involves only people who actually have the disease of concern, screening involves taking medical action on many people, few of whom will actually have the disease being screened for. So benefits are realized for only those few, while costs are incurred for the entire screened population, and harm may result for many whom the screening does not benefit.

---

[49]   Welch, H. G., *Less Medicine, More Health: 7 Assumptions That Drive Too Much Medical Care* (Beacon Press, 2015), p. 66. See also Welch, H. G. and Fisher, E., Income and cancer diagnosis—When too much care is harmful. 2017. *N. Engl. J. Med.*, 376(23):2208–2209 ("there are reasons to wonder whether wealthier people receive too much care … resulting in overdiagnosis and potentially unnecessary treatment").

Examples of harms from screening include the cascading events in the case described above: four follow-up biopsies, an incidental finding, follow-up surgery, confusion over whether the surgery needed to be redone, "extreme trauma to the patient and a developing panic problem," and follow-up care with a psychologist—not to mention spending many hours and dollars and the associated stresses. Other possible harms include complications (including botched execution) from whatever procedures the patient undergoes, and new health risks such as increased radiation exposure from certain radiological scans. Dr. Welch sums up the situation as follows:

> … the odds are stacked against screening: many (often thousands) must be tested, to potentially benefit only a few. Any harms from the testing process … are multiplied since so many people are going though it.

> …Think of what needs to happen for a screening test to even have a chance of working for the few it could help. Everyone must be made concerned enough to get screened, everyone must be tested, many re-tested and needlessly alarmed, while others are over-diagnosed and overtreated.

> In other words, it's the harms that are certain, not the benefits. [50]

From this perspective, it is unsurprising that, as Dr. Welch observes, "only one cancer screening test … has definitively been proven to help people live longer: lung cancer screening in heavy smokers." In that population, "lung cancer is a big component of their overall death rate." In broader populations, no one disease is a similarly big component of deaths. This reality lessens the value of disease screening in broader populations, when harms, costs, and benefits are all taken into account.

Dr. Welch does not reject screening. Screening for lung cancer, for example, is appropriate even though it is driven by fear as "an integral part of anti-smoking campaigns." His point about screening is that "we need to carefully pick and choose. … it should always be an informed choice." [51]

Screening for unidentified disease is just one example of how medical activity can cause unexpected harms. The output of screening and much other medical activity is information. Dr. Welch's underlying argument is that "more information is not always better." For example, "diagnostic tests can be invaluable … But they can also be confusing, distracting and anxiety provoking. Clinical information is a double-edged sword—and physicians are having to sort through more and more of it." [52]

So the question is how to improve the informational supply chain so that patients and clinicians alike can better sort through more and more information. This sorting should enable distinguishing between useful information (which should be taken into account) and pointless information (which should not be collected in the first place). Making this distinction should not

---

[50] *Less Medicine, More Health*, pp. 65, 72.
[51] Ibid., p. 64.
[52] Ibid., p. 85.

be left to unaided human judgment. Judgment should be informed by external tools used before diagnostic testing or therapeutic action are undertaken.

3.2   MISSED AND MISGUIDED TREATMENTS

Turning from diagnostic to therapeutic decision making, here again misguided reliance on the human mind looms large.

3.2.1   TOXIC OPIOID TREATMENT

Overuse of opioid drugs for pain relief—a major public health issue—is rooted in defective medical training and practice. Dr. Marty Makary wrote of how he was trained about opioids:

> [In] my residency training … I was taught to give every [surgery] patient a boatload of opioid tablets upon discharge. The medical community at large ingrained in all of us that opioids were not addictive and urged us to prescribe generously. And that's exactly what we did. … I was unaware of any national best practices for what should be prescribed to a patient after a standard operation.

Then his father had an operation for which Dr. Makary himself had "customarily prescribed 60 opioid tablets." Yet he watched his father "recover comfortably at home with a single tablet of ibuprofen."

This experience led Dr. Makary to investigate the issue in his own institution (Johns Hopkins). He found that opioid prescribing "is really a matter of surgeon style or preference," that their Epic EHR system "had an e-prescribing default that recommended a 30-day supply" of opioids, that chief residents "yell at" interns who prescribe anything less than 30 pills, that prescriptions are "based on what our last resident taught us," that residents "don't know [the outcomes] because we don't follow the patients after they go home like the nurse practitioners do." When Dr. Makary asked a nurse practitioner who "calls every patient at home after surgery," she informed him, "Marty, they don't need any opioids."[53]

Given this chaotic non-system of "care," it's no surprise that prescribing practices became toxic and opioid drug vendors became predatory. Primary elements of this situation include the following.

- Medical decisions lacked scientific basis, but rather were founded on a delusion that "opioids were not addictive."

- No feedback system existed for testing this delusion and correcting it in practice.

---

[53] The above description comes from Chapter 9 of Dr. Makary's remarkable book, *The Price We Pay: What Broke American Health Care – And How to Fix It* (Bloomsbury Publishing, 2019). See also his earlier book, *Unaccountable: What Hospitals Won't Tell You and How Transparency Can Revolutionize Health Care* (Bloomsbury Publishing, 2012).

- Senior doctors thus had unfettered discretion to prescribe opioids for pain relief "as a matter of surgeon style or preference," in other words, based on whatever happened to be in each surgeon's mind.

- Doctors in training were expected not to challenge but to follow the idiosyncratic practices of whomever they happened to be training under.

- Patients played no role in deciding whether opioid drugs were advisable or effective for their own pain.

Doctors would not have become contributors to the opioid crisis if they functioned within an organized system of accountable care, as described in Part II of this book. Opioid drugs are one of multiple treatment options for pain problems. Deciding among these options should begin with either (a) including current pain as a problem on the problem list, or (b) anticipating pain as a complication to watch for when planning surgery or other procedures expected to cause significant pain. Follow-up action should include careful definition of the patient's pain and the full range of possible treatment options (not just drugs), taking into account the pain's cause, its severity, its impact on the patient's life, and the pros and cons (such as addiction risk) of each pain relief option, all as individually applicable to each patient.

For example, if the pain's cause is a surgical procedure, an opioid prescribing guideline should not be applied without first verifying that no better options are available for the patient at hand. Verifying this requires taking into account such data as the level of expected pain (which is procedure-specific), the effect of the patient's co-morbid conditions, interactions with drugs the patient is already taking, the availability of non-addictive pain relief options, the anticipated effectiveness and side effects of each option, the patient's prior vulnerability to addiction, and the patient's own weighing of relevant trade-offs such as avoiding pain vs. avoiding addiction risk. All these details and more should be systematically elicited in the initial investigation. Only then should alternative options be compared and one chosen.

But clinicians typically do not achieve such thoroughness. This gap between what clinicians do and what patients need is the rule, not the exception. In many cases, lack of thoroughness turns out to be harmless error, or harm occurs but is less visible than with the opioid epidemic. So the "system" gets away with it, sometimes even when a pattern of serious harm becomes visible.

People started waking up to the opioid crisis years ago; yet it continues, in part because traditional medical decision making continues.[54] So "in a baffling number of cases, doctors still ignore [CDC] safety guidelines … a lot of doctors still get almost everything wrong when treating

---

[54] *Overdose Deaths Accelerating During COVID-19*, CDC Release, Dec. 17, 2020, stating: "the highest number of overdose deaths ever recorded in a 12-month period," 81,000, occurred in the year ending May 2020. CDC's release was accompanied by a Health Advisory recommending various measures to cope with the problem. These measures can be regarded in part as work-arounds to compensate for failings in medical practice.

pain," according to a recent NPR report.[55] This does not refer to "crooked doctors running illegal pill mills." On the contrary, "a growing number of studies show that was never the real problem." According to an emergency room doctor at Stanford who studies prescribing patterns, "It's not just a handful of doctors doing it, we kind of all are. It's really become a part of our culture that this is normal." At the same time, opioids are *under-prescribed* for patients with sickle-cell disease and cancer, as stated in a new CDC study of opioid prescribing for various specific conditions.[56]

In response to this "normal" but dysfunctional culture, some organizations have followed the established approach to altering doctor habits—clinical practice guidelines. Whether in the form of text documents, EHR components, or CDS tools, guidelines are intended to inform doctors of "evidence-based" guidance for decision making. But guidelines may be subject to improper influences and outright corruption.[57]

That problem aside, practice guidelines have had limited effect in improving decisions generally (see Section 3.3.1) and opioid-prescribing in particular. The NPR report quotes the lead author of the new CDC study (see Notes 54 and 55): "It's possible that some clinicians just simply aren't aware of existing evidence-based recommendations. The other possibility is that they are aware and they just choose not to follow them." This suggests that more awareness or more enforcement or both are needed. But the new CDC study offers no clear remedy. Referring to inconsistency

---

[55]   Medical Professionals Still Prescribing Dangerously High Amounts Of Opioids In U.S., July 17, 2020.

[56]   Mikosz, C., et al., Indication-specific opioid prescribing for U.S. patients with medicaid or private insurance, 2017. *JAMA Netw. Open.* 2020;3(5):e204514 (May 11, 2020). DOI: 10.1001/jamanetworkopen.2020.4514. See page 2, citing three reports of "undertreatment or delay in pain treatment in instances in which the benefits of opioids might outweigh the risks."

[57]   The EHR vendor Practice Fusion recently paid $145 million to resolve criminal and civil investigations relating to its EHR software. According to a U.S. Department of Justice announcement, the company "admits that it solicited and received kickbacks from a major opioid company in exchange for utilizing its EHR software to influence physician prescribing of opioid pain medications." (Beyond this corruption of guidelines, the company also "accepted kickbacks from the opioid company and other pharmaceutical companies and also caused its users to submit false claims for federal incentive payments by misrepresenting the capabilities of its EHR software.") Another example is a U.S. District Court decision against a subsidiary of the nation's largest managed care company, *Wit v. United Behavioral Health* (UBH), which involved practice guidelines internally developed at UBH for cases of substance abuse and other mental health problems. The court found that "UBH's cost-cutting focus 'tainted the process, causing UBH to make decisions about Guidelines based as much or more on its own bottom line as on the interests of the plan members, to whom it owes a fiduciary duty.'" Klepper B., When Health Care Organizations Are Fundamentally Dishonest (The Health Care Blog, Mar. 19, 2019), quoting the court decision. This blog post goes on to point out how this cost-cutting pressure applied against health plan beneficiaries occurs at the same time that "U.S. health care is awash in excess" with "egregious unit pricing and unnecessary services."

Guidelines can also be distorted in favor of coverage, because guideline authors could "prefer to avoid [recommendations that] may result in payers denying coverage. They may also have a vested interest in a particular diagnostic or therapeutic approach." Or a guideline's recommendation for shared decision making "may mislead users, rather than guiding them, by including options that are unlikely to yield benefit that patients will value." Rabi, D., Kunneman, M., and Montori, V., When guidelines recommend shared decision-making. *JAMA* 2020;323(14):1345-1346.

between prescribing patterns and guidelines, the study lamely concludes: *"Implementation guidance that emphasizes evidence-based recommendations* has the potential to better align opioid prescribing practices with evidence … [emphasis added]."* It's not clear what this means. Does "implementation guidance" refer to something different than traditional practice guidelines? Or is this language distinguishing between guidance that "emphasizes evidence-based recommendations" and guidance that does not?

The crisis and the remedies are not specific to opioid use and pain relief. Instead, the crisis is symptomatic of disorganized, unaccountable medical practice. Remedies are needed for medical practice across the board. That may sound overly ambitious, but a broad-based approach would be more feasible, more powerful, and more cost-effective than addressing various crises one by one.

3.2.2 HER PARENTS FINALLY BELIEVED HER, BUT HER DOCTORS DID NOT

In diagnostic odysseys such as those described in Section 3.1, doctors often dismiss what patients try to tell them. This also happens after diagnosis and during treatment (which may itself involve misdiagnosis of continuing symptoms). Here is an appalling example.

> When eight-year-old Elizabeth told her parents that her kidney cancer, which had been in remission, was back, her parents believed her. She was experiencing severe pain in her legs that only morphine could relieve. But the little girl's doctors dismissed her complaints as psychological, stemming from her fear of the cancer recurring. Though Elizabeth had been scheduled for an MRI, the test was canceled because the doctors thought it wasn't necessary. As her mother, Leila, recalled, "The doctors said that my daughter was being manipulative and that I was only raising her anxiety by continuing to take her to doctors." Leila did everything she could not to reinforce Elizabeth's complaints about pain, believing the doctors' view that all was well and that her daughter's cancer phobia needed to be overcome, not encouraged. Leila recalls with a profound sadness how she insisted one Christmas morning that her daughter crawl down the stairs to open presents despite her cries for relief from her pain.
>
> Over the next three months, Elizabeth's symptoms worsened, and she lost almost twenty pounds, going from fifty-eight to forty pounds. Her mom called the doctors about twenty more times, and they continued to assure her that her daughter's troubles were in her head. Elizabeth continued to worsen and became totally withdrawn, not talking to anyone; her doctor had her placed in an outpatient psychiatric ward. Meanwhile, "She just laid there with her eyes closed," her mom says.

With the instinct that only a parent has, Leila called her own mother and said, "Her soul is dying. There is no life in her eyes." She was right. No one knew—not yet—that Elizabeth's brain was swelling because a tumor had metastasized to her spine and was in her brain, pressing on it. On the morning of the day she was finally diagnosed, her dad asked her what she was thinking about, and she replied, "I'm thinking about heaven. There's no pain there."

Later that day she had a seizure. Her doctors ordered an MRI, which proved what Elizabeth had known in her heart all along. The cancer had spread to her spine and brain. Nine months of chemotherapy and a stem cell transplant saved Elizabeth's life, but the delay in diagnosis resulted in permanent paralysis from the waist down.[58]

This case shows doctors' denial of a patient's subjective experience of symptoms, parents' denial of their doubts about the doctors, and the whole system's denial to everyone of a rational approach to care and communication. We return to the issue of the patient's subjective experience in Section 9.1.2.2, where we discuss progress notes in health records.

3.3   MISGUIDED RELIANCE ON ALTERNATIVES TO THE HUMAN MIND

3.3.1   PRACTICE GUIDELINES AND CLINICAL DECISION SUPPORT TOOLS

When driving, we may use maps or other navigation devices as needed. When driving without those aids in unfamiliar territory, we may stop to ask someone for directions, in the hope that whoever we encounter knows the area. That's how medicine is often practiced—asking someone for directions. This is a coping mechanism for not having easily usable maps to the medical landscape. For example, doctors refer cases to specialist consultants, or ask personal colleagues informally, or reach out to the medical community via listservs, blogs, and social media. In doing so, they are calling upon the minds of others, as distinguished from drawing upon knowledge resources directly. On this distinction, see Section 6.1, discussing a concept from Karl Popper.

This reliance on the minds of others is the case even though knowledge resources abound, in the form of medical journals, textbooks, and other resources available in paper and electronic form. But these are not easily usable during patient care, unlike maps and navigation devices during travel. So medicine developed practice guidelines and protocols[59] (which we refer to collectively as guide-

---

[58] *Wall of Silence: The Untold Story of the Medical Mistakes That Kill and Injure Millions of Americans* (Note 34), pp. 22–23.

[59] Protocols are commonly understood to impose rigid, step-by-step requirements, as distinguished from more flexible guidance.

lines), sometimes authored by experts in the specific subject matter. These are designed to synthesize and distill knowledge into consensus recommendations. But these guidelines, while sometimes an improvement on the status quo, have not had the intended effects. For one thing, there are too many of them, as summarized by a 2014 article discussing guideline use in the UK:

> …the number of clinical guidelines is now both unmanageable and unfathomable. One 2005 audit of a 24-hour medical take [sic] in an acute hospital, for example, included 18 patients with 44 diagnoses and identified 3,679 pages of national guidelines (an estimated 122 hours of reading) relevant to their immediate care.[60]

This proliferation of guidelines simply reproduces information overload. Moreover, because they are intended to follow "evidence-based medicine" (EBM), guidelines reproduce the failings of EBM that many critics have identified. These include the following.

- EBM is subject to bias and distortions, especially from industry influence.

- EBM recommendations are typically based on statistically significant benefits determined from the artificial conditions of clinical trials. Those benefits are often unlikely to be achievable in real-world practice conditions and populations.

- EBM recommendations often rely on algorithmic rules. By their nature, these rules take into account only a few features of a problem situation, excluding important details that must be considered for actual patients.

- EBM recommendations are typically not designed for patients with multiple complex conditions; yet such patients are the norm with chronic disease, especially in the elderly, which is the population with the highest costs of care.

Guidelines invariably acknowledge that applying them to individual patients requires "clinical judgment." But that term is a euphemism for the idiosyncratic, limited cognitions of each fallible clinician. To protect against that fallibility, guidelines may attempt to lay out all the details and factors that individualized decisions must take into account. But that attempt makes guideline documents too long and cumbersome to readily use under real-world time pressures. The complexity increases still more when there is a multiplicity of inconsistent guidelines to sort through.

Beyond the failings of EBM as a source of guidance, the most fundamental difficulty with traditional practice guidelines is that they provide only general knowledge. They are not a tool for *applying* knowledge to patient data. So they don't enable navigating the medical landscape in real time in the way that maps and navigation devices do. That capability is still left to the unaided

[60] Greenhalgh, T., Howick, J., and Maskrey, N., Evidence-based medicine renaissance group. Evidence based medicine: A movement in crisis? *BMJ* 2014 Jun 13;348:g3725. DOI: https://doi.org/10.1136/bmj.g3725, citing Allen, D. and Harkins, K., Too much guidance? *Lancet* 2005;365:1768.

human mind. For the mind, applying general knowledge to individual patients is an intricate and error-prone process. Moreover, it depends on personal knowledge and analytic abilities that vary from one person to another.

Recognizing these points, clinicians and health IT specialists have for decades been trying to develop electronic clinical decision support (CDS) tools. These are designed to implement practice guidelines as usable tools in patient care. But the tools have been disappointing in practice: "clinical users' experiences with implemented CDS have been sub-optimal, and CDS development and deployment remain inconsistent and fraught with re-work, even among HHS agencies."[61] A primary reason for this situation is that CDS tools are often designed to guide the user toward the same recommendations stated by traditional written practice guidelines.

Traditional guidelines are misconceived. In making recommendations (favored treatment alternatives, for example), they attempt to prescribe optimal decision choices, when what is needed are optimal decision processes. Optimal processes lead to individualized choices, each one optimal for each patient.

This goal requires a different medium than ordinary text, written or electronic. The different medium is CDS tools designed to couple vast medical knowledge with detailed patient data and generate actionable output. The output must be organized for decision making purposes and precisely tailored to the patient's individual problem situation. See Section 8.3.

3.3.2   INTERNET SEARCH ENGINES

At first glance one might think that Internet searches are sufficient as a CDS tool to aid the clinician's mind. As illustrated by the second Addison's disease case discussed in Section 3.1.3.2, an undiagnosed problem can be entered into search engine such as Google.[62] The search results can suggest diagnostic possibilities and related findings that the clinician might not have thought of. That may well be an improvement on the status quo. But such use of Internet searches has three serious limitations.

First, the search results depend on what terms the user chooses to enter in the search engine. That initial exercise of judgment introduces the cognitive vulnerabilities that external tools should protect against. Inevitably, clinicians will differ in the initial findings they make and what they choose to enter in the search engine. Different combinations of findings entered initially may yield search results pointing in very different directions. The patient thus has no assurance that the clinician is following an established best practice for investigating the problem.

---

[61] From Clinical Decision Support (CDS) Workshop" (9/15/20) by Office of the National Coordinator for Health Information Technology (ONC).

[62] Tang, H. and Ng, J. Googling for a diagnosis—use of Google as a diagnostic aid: internet based study. *BMJ*, DOI: 10.1136/bmj.39003.640567.AE (published November 10, 2006).

Second, the repository for Internet searches is an unfiltered, unstructured body of information—the entire world wide web. Instead, the repository to search on should be a distilled body of information selected and structured in advance for precise relevance to the problem being investigated. For example, if the problem is diagnosis of acute abdominal pain, then the repository should be a carefully selected body of authoritative guidance on (i) the full range of possible diagnoses for acute abdominal pain, and (ii) the initial needed findings on the patient for determining which of those possible diagnoses are worth considering.

Third, the output of an Internet search is determined by algorithms of limited transparency that crudely filter and structure the search results (typically by first displaying results that include every search term). Instead, what the patient and clinician need is precisely relevant knowledge (possible diagnoses worth considering for that patient) coupled with actual patient data on each possibility, plus guidance on what it all means. This output should be organized as options (diagnostic possibilities) and evidence for and against each one. Such output is hardly what one finds in Internet search results.

Accordingly, the cited *BMJ* article on use of Google as a diagnostic aid acknowledged serious limitations in that approach.

> We suspect that using Google to search for a diagnosis is likely to be more effective for conditions with unique symptoms and signs that can easily be used as search terms … Searches are less likely to be successful in complex diseases with non-specific symptoms … or common diseases with rare presentations … . The efficiency of the search and the usefulness of the retrieved information also depend on the searchers' knowledge base.

The Internet has created an unprecedented excess of available information. This new excess magnifies a dilemma already inherent in any attempt to apply complex general knowledge to specific problem situations. This dilemma requires more than a search engine alone can provide. Coping with this dilemma requires: (1) careful distillation and arrangement of potentially useful knowledge in a specialized repository; and (2) a tool for linking that distilled knowledge with carefully selected data about specific problem situations and organizing the results in maximally usable form, while filtering out whatever is truly extraneous. That approach is discussed in Chapter 8.

3.3.3   THE WISDOM OF CROWDS? MISGUIDED TOOLS FOR CROWDSOURCING AND COLLECTIVE INTELLIGENCE

Now let's further consider the status quo. Doctors access the subjective knowledge of others via informal consultation with colleagues, as described by Dr. Lisa Sanders in her recent bestselling book on diagnosis. "In medicine, doctors accept that no one knows everything. Our knowledge is shaped by experience, training and personal interest. We all reach out to our community of doctors when we are stumped." Now "the Internet offers … a broader community," reachable by blog and listserv posts and social media.[63]

This informal consultation is inherently limited and random. It's far better than working in isolation, but it falls far short of a tool-driven system for:

1. *accessing objective knowledge relevant to the problem*, as distinguished from whatever subjective knowledge clinician colleagues happen to provide;[64]

2. *identifying the significant details of the patient's problem to take into account*, as distinguished from whatever details an inquiring clinician happens to submit to colleagues; and

3. *reliably linking the details with applicable knowledge and organizing the results*, as distinguished from the analyses received from colleagues, which are likely to be idiosyncratic, incomplete, based more on judgment than documented analysis, and not vetted, consolidated, and organized to aid the inquiring clinician or the patient.

These same three failings apply to formalized collective intelligence, as discussed next.

After informal "reaching out" to colleagues, the next step is a formal referral process. Typically, a primary care doctor refers a difficult case to a specialist for a second opinion (although patients themselves sometimes seek out a second opinion). The referring doctor often does not attempt a final diagnosis, instead simply asking for help with preliminary diagnosis of an unexplained problem. Referrals often occur at the diagnostic stage, but specialist consultants may advise on treatment as well as diagnosis. Often these referrals involve nothing more than a medical record review by the consulting specialist. Sometimes the specialist has a face-to-face encounter with the patient and may collect additional data.[65]

The formal referral process raises a number of questions. What reason is there to believe that the second doctor is correct? What might a third doctor find? (The cited Mayo Clinic study acknowledged the lack of longitudinal follow-up to confirm the second opinion.) How does the first doctor know which specialty should be consulted? (Multiple specialties may seem relevant as various symptoms appear.) Given that patient problems cross specialty boundaries, of what value are the conclusions of specialists, who tend to have tunnel vision based on their own niches? (Recall

---

[63]   Sanders, L., *The heat of the night*, *NYT Mag.*, Sep. 10, 2010. Reprinted as "The Flu That Stayed" in *Diagnosis: Solving the Most Baffling Medical Mysteries* (New York: Broadway Books, 2019) (see p. 12).

[64]   As to objective and subjective knowledge, see the discussion of Karl Popper in Section 6.1.

[65]   See this Mayo Clinic study of its referral practices, which are not limited to medical record review. Van Such, M. et al., Extent of diagnostic agreement among medical referrals. *J. Eval. Clin. Pract.*, 2017. 23(4): pp. 870–74. DOI: 10.1111/jep.12747. This study found that in almost 300 consecutive referrals, 12% of diagnoses by the referring doctor agreed with the diagnoses by the second opinion doctor, 66% of the initial diagnoses were better defined or refined in the second opinion, and 21% of the initial and second opinion diagnoses were "distinctly different." In these three groups, the mean cost per referral was $1,288, $1,794, and $4,767, respectively.

the Parkinson's disease case discussed in Section 3.1.2.) Why are formal referrals so often inconclusive, with patients wandering from one specialist to another before a solution is found, if ever?

In short, the formal referral process is only marginally different than informal consultations. Both forms of collective intelligence are subject to the three failings listed above.

Compared to these traditional alternatives, more comprehensive, systematic approaches to collective intelligence have been developing. These involve virtual, peer-to-peer second opinion networks, open to clinicians generally. A prominent example is Medscape Consult, the subject of a recent study.[66] The broad reach of these networks makes them an advance on traditional practices. But their output is still rooted in subjective opinions, and they are still subject to the three failings listed above. These three failings are implicitly acknowledged in the cited article on Medscape Consult.

- As to "*accessing all knowledge objectively relevant to the problem*" (the first failing listed above), the study acknowledges that the consulting opinions "are only based on potentially biased information" from the inquiring clinician. Moreover, biased or not, the information is necessarily selective and dependent on the clinician's personal understanding. See the two examples of case presentations in the article supplement. Moreover, the universe of clinicians who happen to see the inquiry and choose to respond is necessarily limited and somewhat random, as distinguished from systematically connecting with authoritative sources of relevant expertise via the medical literature or otherwise.

- As to "*identifying the significant details of the problem to take into account*," the study observes: "the responders are not afforded the opportunity to fully examine the patient. Thus, important aspects of the physical exam, interpersonal discussion, and subtle clues to diagnosis are likely lost in this digital space, which could potentially increase the risk of diagnostic inaccuracy" (PDF p. 4). This failing is inherent in any referral to another clinician who works secondhand, based on selective details. The only protection against this failing is to equip the initial, referring clinician (plus the patient) with the relevant external knowledge, rather than outsourcing the knowledge retrieval and analysis to some other clinician. In that way, the relevant knowledge is delivered to the parties (referring clinician and patient) best positioned to apply that knowledge effectively to the details of the case. If they still need help with solving the problem, they may then wish to consult an outside clinician, but they can do so by providing much better information than is normally the case.

---

[66] Muse, E. et al., From second to hundredth opinion in medicine: A global consultation platform for physicians. *NPJ Dig. Med.* (2018) 1:55; DOI: 10.1038/s41746-018-0064-y. A supplement to this article provides two sample case presentations and multiple responses to each.

- As to "*reliably matching the details with applicable knowledge and organizing the results,*" see the responses elicited by Medscape Consult shown in the article's supplement (four responses to one case and 27 responses to the other). The responses were brief comments of the sort that might be received in a "curbside consult"; none were detailed analyses of the sort received in a formal referral. Some of the responses may have been quite useful, but they do not systematically lay out options to consider with evidence for and against each option. Instead, they consist of what appear to be off-the-cuff opinions and suggestions, without citations to the literature.

Another prominent example of the collective intelligence approach is the Human Diagnosis Project (Human Dx). This project has built a sophisticated Web-based tool that enables clinicians to submit their cases to other clinicians world-wide and receive back case-specific "collective intelligence."[67]

The Human Dx tool undoubtedly improves on ordinary practice. It stands to reason that "two heads are better than one," that collective thinking is better than the thinking of isolated minds. But that goal—merely to exceed individual doctor performance—sets the bar too low.

Both individual and collective performance are inevitably compromised by the human mind's limited ability to cope with massively detailed knowledge and data. It stands to reason that human minds, individually and collectively, do better when they use external information tools. Human Dx has built such a tool, but its design incorporates, rather than bypasses, important failings of the human mind.

The Human Dx tool operates by

> …combining the collective intelligence of medical professionals and trainees with machine learning. … Using the Human Dx system, medical professionals and trainees can collaborate on any case, question, or other medical topic. As they do, *the system encodes their thought processes and decisions* into structured clinical data to map the steps to help any patient [emphasis added].

Note the approach of encoding clinicians' "thought processes and decisions."[68] This approach seeks to "model the processes of how a doctor determines a medical diagnosis when exposed to incomplete information."[69] For example, the online tutorial explains (at 0:41), "Like a real case in practice, only the chief complaint is revealed at the start. Based on that, I enter my initial

---

[67]   For a general discussion of the collective intelligence approach, see Kurvers, R. H. J. M., Herzog, S. M., Hertwig, R., et al., Boosting medical diagnostics by pooling independent judgments. *Proc. Natl. Acad. Sci. U.S.*, 2016;113:8777–8782. See also the links to the three medical journal articles at humandx.org/, where the project states that its system has been "extensively validated … with leading research institutions."

[68]   From Project Background, Q&A 1.1, What is the Human Diagnosis Project?

[69]   From Q&A 2.7 (6th bullet), What are existing computerized diagnostic solutions and how do they compare to the Project?

differential diagnosis," meaning that the user enters a list of possible diagnoses to consider, based on personal judgment.

That up-front exercise of judgment is inherently limited, idiosyncratic, and vulnerable to countless missteps. The same failings apply to what comes next after the differential: judging how to investigate whatever diagnostic possibilities the clinician happens to have judged relevant.

A completely different approach is possible. This approach elicits detailed, carefully selected information up front, *before* the clinician exercises judgment. In this way precisely relevant information is assembled and organized, thereby protecting against judgment's vulnerabilities. But this approach also preserves freedom for judgment to operate after the informational foundation is laid. See Chapter 8.

Human Dx is founded on a misplaced faith in the collective intelligence of practitioners. It's true that intelligence is likely to be improved by operating collectively rather than individually. But the human mind, individual or collective, is simply not a good vehicle for assembling the detailed information (patient data and medical knowledge) required for trustworthy decision making. That informational foundation should be laid by external tools designed to optimize the basis for decisions. These tools must be employed *before* applying human intelligence and judgment, individual or collective. Then human and artificial intelligence may have a role to play in further improving the informational basis for decisions.

As a tool for aggregating practitioner judgments and generating feedback, the Human Dx project is interesting and impressive in many ways. It has a worthy goal—to release information that "is siloed in medical research, academic handbooks, disparate closed systems, and the minds of individual practitioners" (Q&A 1.2). We are interested in the possibility that the project might be recast and adapted to serve as a vehicle for practitioners and patients to provide feedback on medical knowledge as described above. But we disagree with how and where the project keeps the human mind in the loop.

Rather than allowing judgment to operate as soon as the chief complaint is entered, a completely different approach would be to *defer* judgment until after thorough data collection and analysis by external tools. We describe that approach in Chapter 8.

3.3.4   ARTIFICIAL INTELLIGENCE

At first glance, AI tools promise to bridge the gap between the human mind's limited capacities and the enormous complexities of medical decision making. But that promise is uncertain. Both human and artificial intelligence have their own capacities and vulnerabilities. Reliance on each form of intelligence must be selective and carefully matched to the problem situation. The necessary matching requires something like the tools described in Part II. (It may turn out that AI itself will become an additional tool for the necessary matching.)

In developing and using AI, we must balance between two risks. On the one hand, we risk uncritical acceptance, thus not recognizing AI's potential for harm until too late (see our discussion of "the alignment problem" in Section 3.3.4.4). On the other hand, we risk rejecting AI indiscriminately out of ignorance and fear, thus missing out on its enormous potential for good. Both risks are somewhat analogous to those we face with vaccination. In that arena, meticulous scientific practices are essential to develop effective vaccines and guard against their risks. The same is true with AI in medicine.

To explore the above issues, this section describes recent developments with AI in medicine and the new capabilities of machine learning.

3.3.4.1 Distinction Between Traditional Software and Advanced Artificial Intelligence

The tools described in Part II are traditional software. They do not themselves employ advanced AI. But they help enable targeted use of specific AI tools, including those based on machine learning, and integrate those tools with existing patient data, medical knowledge and human intelligence. To explain these points, we first need to distinguish AI from traditional software.

The term "artificial intelligence" (AI) refers to advanced software with completely new techniques and capabilities not available with traditional software. The tasks associated with traditional software and advanced AI, respectively, can be roughly described as follows.

- Deterministic tasks, such as sorting, mathematical calculation, linking related database items, and other tasks executable as explicit, step-by-step, computer-codable instructions, exhaustively performed. Tools for these deterministic tasks we refer to as traditional software.

- Indeterminate tasks, such as vision and image analysis, natural language processing (NLP), vehicle autonomous driving, and other functions that cannot be translated into step-by-step instructions executable by traditional software. Tools for these indeterminate tasks we refer to as advanced AI. The most powerful advanced AI usually employs "machine learning," that is, deriving feedback from experience to improve performance, without explicit, programmed instructions.

Advanced AI is at the cutting edge of computer science. It is designed for problems that humans can handle only within limits (vision and NLP, for example) or cannot handle at all (optimization of ICU activities by continuously adjusting more variables than humans can process, for example). These problems are indeterminate for various reasons—typically because information is incomplete, or information is too complex for traditional software instructions to handle, or such instructions would be either not definable or prohibitively time-consuming to execute for available computing power.[70]

Traditional software, because it handles deterministic tasks, is well-defined, predictable, and transparent in its operations. These characteristics limit its scope but give it reliability and trustworthiness. Human intelligence can handle a far greater scope of tasks than traditional software, but without the same reliability and trustworthiness.

Advanced AI has the power to handle certain tasks beyond the capabilities of either traditional software or human intelligence. But advanced AI usually involves a trade-off between power and transparency. That trade-off is acceptable in narrow niches involving certain kinds of pattern recognition where narrow AI is clearly superior to human intelligence and lack of transparency creates little risk. In other niches, narrow AI involves significant risk, so that greater transparency (and thus risk mitigation) is needed before AI tools will be tolerated (a non-medical example is AI for autonomous driving).

As yet, there is no form of artificial general intelligence (AGI) comparable to the general intelligence of humans, although research on AGI is being actively pursued. Without AGI, there is little reason to expect AI for general diagnostic or therapeutic decision making outside of narrow niches.

3.3.4.2 AI and the Informational Supply Chain in Medicine

Recall the two stages of medical decision making as presented in Section 2.1. The first stage is assembly of information. The second stage is making the ultimate decision in light of the assembled information, based on the decision maker's values and judgment.

Assembling information in the first stage depends on the informational supply chain. Traditionally, the doctor's trained intelligence has been the key link in that chain. This central role for human intelligence, as discussed in Section 2.1, is unworkable. The alternative we advocate is the following.

- The informational supply chain should rely primarily on traditional software tools with the core functionality described in Chapters 8 and 9.

- The role for intelligence, whether human and artificial, is to *supplement* these traditional software tools.

The supplemental role means that intelligence should be permitted to add to, but not subtract from, information supplied by traditional software tools. Stated differently, traditional software tools should be designed and used to set a high bar in accuracy, completeness, and objectivity for the information decision makers take into account. Then human and artificial intelligence should pro-

---

[70] See, generally, Christian, B. and Griffiths, T., *Algorithms to Live By: The Computer Science of Human Decisions* (Henry Holt 2016), p. 4.

vide inputs in specific contexts where they surpass that high bar. This approach leads to evolutionary, continuous improvement of the information on which medical decisions depend. See Section 6.1, discussing Karl Popper's evolutionary concept of knowledge, enabled by moving knowledge from the human mind to books and other external devices.[71]

To illustrate the supplemental role in the first stage of decision making:

- for human intelligence, a simple example of this supplemental role is entering free text to annotate structured or coded data entries. Section 8.3.3.4 discusses this example in more detail; and

- for artificial intelligence, examples of this supplemental role involve pattern recognition of various types. In the diagnostic area, these include, as summarized by Dr. Topol, "interpretation of medical scans, pathology slides, electrocardiograms, or voice and speech." In these narrow diagnostic contexts (as distinguished from "whole patient diagnosis"[72]), advanced AI has clearly surpassed human abilities. Similar potential exists for AI in some therapeutic contexts.[73]

Uses of AI for narrow pattern recognition are essential where they are clearly more reliable and trustworthy than human sense perceptions and cognitions.

Suppose general AI tools emerge—tools that seem capable of "whole patient" decision making. The question will become whether such tools should just play a supportive, assistive role in the first stage of decision making, or whether they should instead supersede the human mind in both stages. The latter, more autonomous role is envisioned by some who see advanced AI taking over large swathes of human decision making in many fields, including medicine.

AI is at an early stage, and its role will no doubt expand dramatically in many areas, including medicine. But we don't believe that AI or any external tool should ultimately remove the human mind from either of the two stages of medical decision making. In the first stage (assembling information), human intelligence should always have a supplemental role, because that role is essential for feedback and continuous improvement. In the second stage (making ultimate decisions), human intelligence should have the primary role, because retaining that role is essential to protect human values.

These principles mean that the normal role for CDS tools (whether traditional software or AI) should be only to present options and evidence for decisions, without determining the tool user's second-stage choice among the options. This point about the second stage, however, needs to be qualified. We later distinguish between information-driven decisions and values-driven/prefer-

---

[71]   For further discussion, see *Medicine in Denial* (see Note 4), pp. 45–46, 48, 102, 108–110, 211.
[72]   *Deep Medicine* (see Note 7), pp. 56–57.
[73]   Liu, S., et al., Reinforcement Learning for Clinical Decision Support in Critical Care: Comprehensive Review. *J. Med. Internet. Res.*, 2020;22(7):e18477.

ence-sensitive decisions (see Section 8.1). For information-driven decisions where rational prefer-ences do not vary once people are fully informed, it may be legitimate for policymakers rather than individuals to make the ultimate decision, thereby protecting individuals from being manipulated by third parties (or their own vulnerabilities) into making irrational choices. In that situation, the ultimate choice is made by policymakers, and not by patient or clinician users of the CDS tool. The tool's role is to elicit the relevant information and inform the user on how the chosen policy applies to the patient in light of the information elicited.

An example is opioid drugs as a therapeutic option for pain relief (see Section 3.2.1 for background). The CDS tool would elicit all relevant details about the context (such as any prior addiction and other medical history indicating unusual risk for the patient), alternatives for pain relief, and current legal restrictions on opioid alternatives, given the patient's personal risk/benefit context.[74] The CDS tool would display this information to the user. This CDS tool output should be available to the patient and any clinician or pharmacist the patient might seek out. They should be required, as a standard of care, to review this CDS output and act accordingly.

It is premature to assess these issues in depth. Medicine needs to work out the optimal divi-sion of intellectual labor among traditional software, advanced AI as it develops, and the minds of the patients and clinicians involved. Working out that division of labor should include re-thinking educational and licensure arrangements that tend to lock in current clinician roles. See Chapter 10.

3.3.4.3 Evolving Experiences with AI in Medicine in Comparison with Traditional Software

In the next section we discuss pitfalls of advanced AI. Here we discuss prior experience with more basic AI in medicine. This prior experience shows why using external tools, whether traditional software or AI, requires sound standards of care for managing health information.

Dr. Topol describes[75] a failed attempt at the MD Anderson Cancer Center to use IBM's Watson software. At the time Watson was a leading early AI tool for NLP. Watson was developed in part by being designed to compete against champions of the TV game show *Jeopardy!* But, Dr. Topol explains, "it turns out all Watson did to beat humans in the game show was essentially to ingest *Wikipedia*, from which more than 95% of the show's questions were sourced." IBM mar-keting and the media heavily publicized Watson's success in solving this contrived problem. This publicity led people to believe that Watson's NLP capability would enable ingesting the massive medical literature plus EHR data and thus become able to generate treatment recommendations for cancer patients. But this project failed, even at the threshold tasks of ingesting the medical literature and EHRs.

---

[74] See "What new opioid laws mean for pain relief," Harvard Health Letter (Oct. 2018; updated April 2020).
[75] *Deep Medicine* (see Note 7), pp. 138–39, 156–57. See also Regalado, A., Facing doubters, IBM expands plans for Watson, *MIT Tech. Rev.*, Jan. 9, 2014 (describing plans for Watson to go beyond NLP to a broader set of "cognitive computing" capabilities).

Watson and the medical literature: Dr. Topol notes that two million peer-reviewed bio-medicine articles are published annually. That massive volume suggests that AI would be useful if it could somehow condense and structure the literature to aid decision making. But Watson merely ingested article abstracts—unstructured free text summaries. This alone is useless. "Gleaning information from biomedical literature is not like making sense of Wikipedia entries," Dr. Topol explains. "*A computer reading scientific papers requires human oversight* to pick out key words and findings. … No software today has the natural language processing capabilities to achieve this vital function. But it is certainly coming along."[76] [Emphasis added.]

So it seems that NLP is progressing toward automation of the "human oversight" needed to glean useful knowledge from the literature. Describing this goal, Dr. Topol explains: "At some point in the years ahead, Watson might live up to its hype, enabling all doctors to *keep up with the medical literature relevant to their practice, provided it was optimally filtered* and user friendly." Later, he reiterates the need for "the ability to *ingest all the medical literature… bringing a vast knowledge base to the point of care* of individual patients … facilitating medical diagnoses and optimal treatment recommendations." But these statements obscure key issues.

- Why ingest "all" the medical literature? Much of it is obsolete or inconsistent or du-plicative. Only a small portion can be considered current, authoritative, trustworthy guidance. Moreover, "bringing a vast knowledge base to the point of care" burdens doctors with sorting through all that knowledge. To cope with these pitfalls, text-based practice guidelines were developed long ago to filter and distill the literature. But guidelines have proven inadequate, as discussed in Section 3.3.1. So the issue becomes whether the purposes of traditional guidelines can be better fulfilled by CDS tools in the form of either traditional software or AI.

- What does it mean to enable "all doctors to keep up with the medical literature" as "optimally filtered"? This sounds like using Watson to load the doctor's mind with filtered medical literature, which the doctor must then comprehend and apply to each patient. That leaves the doctor's mind as an error-prone link in the informational supply chain.

- A far better link in the supply chain would be an external tool to match actionable knowledge from the literature with each patient's data. The traditional software de-scribed in Chapter 8 takes this approach by (i) using knowledge from the literature and guidelines to identify needed patient data for the problem, (ii) matching the collected data points with medical knowledge to determine what the data mean, and (iii) organizing the results into options and evidence for the patient and clinician to

---

[76] *Deep Medicine* (see Note 7), pp. 138–39, 143.

consider, with extraneous information filtered out. All this is readily accomplished before human intervention and *without* AI. So the question becomes whether and where AI can be used to improve on traditional software of this kind. For example, if AI is presented with the results from step (iii), can AI weigh the evidence and decide among the options better than the patient and clinician can? See the distinction between information-driven and values-driven decisions in Section 8.1 and the opioid drug example in the preceding section.

- Dr. Topol implies that doctors need to keep up with only the literature "relevant to their practice." This is doctor-centric, tunnel vision—the opposite of patient-centered care. Patient problems cross boundaries between specialized doctor practices, doctors are not the only clinicians seen by patients, and they need to consider knowledge on SDoH outside of the ordinary medical literature. So all knowledge relevant to each patient, not limited by the normal roles of the clinicians involved, must be taken into account. That is part of what the tools described in Chapters 8 and 9 are designed to do.

Dr. Topol does recognize a deep problem with the medical literature, applicable to practice guidelines as well: the medical knowledge they incorporate is limited and simplistic. Specifically, Dr. Topol criticizes "the medical community's fixation with the average patient, who does not exist. For example, there's no consideration of ancestry and ethnic specificity for lab tests," such as the differences between people of African vs. European ancestry with respect to basic tests such as hemoglobin A1C and serum creatinine. Moreover, he rightly criticizes medical information in the form of binary, normal/abnormal classifications. These result in our missing "plenty of information hidden inside the so-called normal range," and thereby "ignoring rich, granular and continuous data that we could be taking advantage of." For example, a steady decline in a male patient's hemoglobin within the normal range over several years could signal a hidden disease process such as internal bleeding or cancer. An AI deep learning system, he argues, could find such signals within "an individual's comprehensive, seamlessly updated information."[77] But current EHRs are not an adequate source for that information.

Dr. Topol also makes the fundamental point that that medical knowledge about the *average* patient "does not answer the practicing doctor's question: what is the most likely outcome [for] a particular patient."[78] Answering that question becomes feasible if "a physician or an AI system would incorporate all of the individual's data—biological physiological, social, behavioral, environmental—instead of relying on the overall effects at a large cohort level" (p. 143).[79] Again, this brings

---

[77] *Deep Medicine* (see Note 7), p. 144.
[78] *Deep Medicine* (see Note 7), p. 144 (quoting Austin Bradford Hill).
[79] *Deep Medicine* (see Note 7), p. 143.

us to EHRs, where "all of the individual's data" should be accumulated in a highly organized form that captures clinical context.

Watson, EHR data, and decision making: Like the medical literature, medical records "also defy the power of AI tools," Dr. Topol states. He finds that this would not be the case if EHRs "were structured comprehensively, neatly, and compactly. We haven't seen such a product yet." (Dr. Topol has not seen the medical record tool described in Chapter 9.) Not surprisingly, therefore, Watson was unsuccessful at ingesting data from disorganized, real-world EHRs. Nor could Watson apply the literature to those data for decision support. As described by Dr. Topol:

> It turns out that it's not so easy to get a machine to figure out unstructured data, acronyms, shorthand phrases, different writing styles, and human errors. … "You have to put in the literature; you have to put in cases" [citation omitted]. Fragmentary clinical data and lack of evidence in the medical literature made this project of little value. Ultimately, the project with Watson cost $62 million and it collapsed. … The problems that IBM Watson encountered with cancer are representative of its efforts to improve diagnosis across medicine. …[80]

IBM encountered problems with not only diagnostic but therapeutic decision making.[81] Dr. Topol concluded: "The difficulty in *assembly and aggregation of the data* has been underestimated, not just by Watson but by all tech companies getting involved with healthcare" (emphasis added).[82]

By comparison, the traditional software tools described in Chapters 8 and 9 implement powerful approaches to "assembly and aggregation" of information, in two ways: matching knowledge with data, and defining standards of data organization in medical records. Both ways involve tools and standards external to the minds of doctors. Without this informational supply chain, neither human nor artificial intelligence can realize their potential in medical practice.

From that perspective, some uses of AI seem pointless. For example, Dr. Topol lists a dozen companies working on AI algorithms to generate doctor notes from doctor-patient encounters as captured in NLP transcripts of audio recordings. Having "this unstructured conversation synthesized into an office note" presumably would free doctors from typing their notes into unwieldy EHRs, which in turn would enhance face-to-face contact with patients.[83]

---

[80] *Deep Medicine* (see Note 7), pp. 55–56. Despite subsequent efforts to further develop Watson's capabilities since the time of the MD Anderson project, IBM is now exiting its AI business in health care. Weiss T., IBM reportedly looking to sell its unprofitable Watson Health business. *Enterprise AI*, February 26, 2021, https://www.enterpriseai.news/2021/02/25/ibm-reportedly-looking-to-sell-its-unprofitable-watson-health-business. See also Strickland, E.. How IBM Watson overpromised and underdelivered on AI health care, *IEEE Spectrum* (April 2, 2019).

[81] *Deep Medicine* (see Note 7), p. 157, citing Ross, C. and Swetlitz, I., IBM's Watson supercomputer recommended 'unsafe and incorrect' cancer treatments, internal documents show (*STAT News*, July 25, 2018).

[82] *Deep Medicine* (see Note 7), p. 56.

[83] *Deep Medicine* (see Note 7), pp. 141–142.

But the doctor's questions to the patient, and the doctor's thinking about the patient's answers, are too often misguided and incomplete. None of this is fixed by automated generation of the doctor's notes. Such a use for AI illustrates our state of denial about the crippling effects of the human mind on the informational supply chain.

Although some use of AI is pointless, AI tools for narrow pattern recognition are finding powerful uses in medicine, such that AI can and should replace human abilities in some contexts.[84] Such uses are multiplying, so much so that their availability and potential relevance for any individual patient are already more than clinicians can be aware of. Moreover, even when the clinician is aware of a particular AI tool and its applicability to a patient's problem, the clinician and patient need some basis for evaluating the pros and cons of using the tool relative to alternative approaches for the problem situation. For example, if a clinician is using an AI tool to diagnose a patient's skin lesion, and if the patient is non-white, the clinician and patient need to know if the AI takes into account the varying appearances of a given skin condition in various non-white populations, which or may not be included in the image samples used to develop the tool.

Both of these issues—awareness and evaluation—can be readily addressed with traditional software tools like those described in Chapter 8. In this way, the informational supply chain can help match patient needs (problems) with the right AI tools, just as can happen with other advances in medical knowledge and technology. Moreover, the health record tools described in Chapter 9 can provide trustworthy data for both developing and using AI tools.

3.3.4.4 Advanced AI and Machine Learning

We now turn to advanced AI. Before discussing its use in medicine, we provide some further background about its recent evolution.

Within the past decade, the AI field experienced a series of pathbreaking advances in "machine learning." This term refers to software tools capable of learning from experience without

---

[84] Caution is still necessary, however. See the article by Oakden-Rayner, L., "Performance is not outcomes; safety in medical artificial intelligence," in Chang, A., *Intelligence-Based Medicine: Artificial Intelligence and Human Cognition in Clinical Medicine and Healthcare* (Elsevier-Academic Press, 2020), pp. 125-126. This article discusses early experience with 1990s computer vision techniques for screening mammography, referred to as computer-aided diagnosis (CAD). Numerous studies showed that humans combined with CAD outperformed humans alone, based on defined "performance testing." These studies led to FDA approval and then Medicare reimbursement for use of CAD. By 2010, an estimated 74% of mammograms were read by CAD. "But in practice, many radiologists felt that these systems did not work very well, and that using them could be frustrating. This feeling was born out as outcomes-based clinical trials were performed over the following decades." The clinical trials showed no improvement with CAD. Unlike the earlier studies, these clinical trials examined actual outcomes in terms of specificity, sensitivity, biopsy rates, cancer detection rates and recall rates. This article concludes that performance testing criteria as seen in medical AI research papers and regulatory approvals are no substitute for real-world outcome comparisons.

being explicitly programmed to do so. This capability enables computers to learn tasks that (i) humans perform but don't know how to program or (ii) humans can't perform at all.

Conceptual models (mathematical and computational) are used to enable machine learning by digital tools. What exactly is going on within the tools, however, is not entirely ascertainable. Moreover, the tools are able to act autonomously, such that humans and digital tools are just different kinds of agents operating in the same social environment. No longer do the human agents have the oversight and control they are used to.

This evolution has gone so far and so fast that, Brian Christian writes, "there is a growing sense more and more of the world is being turned over, in one way or another, to these mathematical and computational models—they are steadily replacing both human judgment and explicitly programmed software of the more traditional variety."[85] The risk is that models embodied in digital tools may act in ways that dangerously diverge from what we intend. This risk is what Brian Christian calls "the alignment problem"—aligning our intentions and values with what the models learn and (especially) what they do.

How these concerns arise depends in part on what category of machine learning is involved use. Machine learning falls into three general categories.[86]

- **Supervised learning:** a machine is instructed to train itself from a dataset of labeled instances, thus becoming capable of applying the labels correctly to new, unlabeled instances. Supervised learning is often described as task-driven. A medical example is using a dataset on hospitalized pneumonia patients to predict survival of new patients for triage purposes.[87]

- **Unsupervised learning:** a machine is instructed to find hidden patterns or groupings in a dataset of unlabeled instances. The initial instructions and the dataset are the only user input. Unsupervised learning can thus be used to explore and label new data. So unsupervised learning is often described as data-driven. A medical example is identifying cancer subgroups based on their gene expressions. Another example is linking disease risks to genomic biomarkers.

- **Reinforcement learning (RL):** This category is fundamentally different from supervised and unsupervised learning. RL involves a trial and error process with positive and negative reinforcement to guide sequential decisions toward a goal. Specifically, a software agent operating in a data environment is instructed to optimize reward/penalty consequences at each successive decision in furtherance of a goal. As a medical example, an environment could be a dataset of ICU care over time from numerous

---

85 Christian, B., *The Alignment Problem* (W.W. Norton, 2020), pp. 11–12.
86 See generally ibid, p. 11, and Chang, *Intelligence-Based Medicine* (Note 84), pp. 83–102.
87 Ibid., p. 82.

devices monitoring various physiological parameters and medical interventions for ICU patients. ICU care requires frequent adjustment of treatment plans in response to feedback from the monitoring devices in order to maximize the likelihood of survival. So RL would involve defining a reward function that would connect treatment decisions with the survival goal. A medical example of RL is optimizing de-intubation timing and sedative dosing in ICUs.[88] In some situations, no basis is known for defining the reward function, so "inverse reinforcement learning" is used to infer a reward function from the dataset where expert clinicians have made the decisions.

The more advanced versions of these three forms of machine learning typically involve "deep learning" from "deep neural networks." A neural network is a computational model with some characteristics of brain neuron networks. A deep neural network involves multiple layers of neural networks, with one layer processing raw data and its output being processed by another layer, and so on for successive layers. Deep neural networks make machine learning dramatically more powerful.

Now we discuss examples of misalignment—various ways in which machine learning can go wrong in medicine. From this perspective, machine learning techniques are analogous to treatment alternatives that may be more or less suitable for an individual's problem situation. As with selecting a treatment alternative, deciding to use a particular machine learning tool should be informed by awareness of its suitability for the individual. That awareness can be provided by the traditional software tools discussed in Chapter 8.

In supervised learning, a primary risk is that the AI tool may be trained on a dataset not representative of the real-world population for which the tool is used. One possible effect is discrimination against minorities. This effect has been recognized in many contexts. A prominent example in medicine occurs with dermatology. AI for visual analysis of skin abnormalities compares a photo of a patient's skin lesion with an image library of diagnosed skin lesions. The problem is that non-white populations are under-represented in dermatology image datasets, compromising accurate diagnosis. Fixing this problem requires a long-term, concerted effort, which is ongoing (see Section 8.4).

Even with a perfectly representative dataset, error may be built into the AI algorithm design. "The problem with machine learning systems is that they are designed precisely to infer hidden correlations in data."[89] Some of those correlations may be meaningless or misleading for the AI tool's intended use. Recall the familiar truism that "correlation does not imply causation." Spurious correlations are a risk regardless of whether an AI tool uses explicit rules or more advanced machine learning.

---

[88] Yu, C., et al., Inverse reinforcement learning for intelligent mechanical ventilation and sedative dosing in intensive care units. *BMC Med. Inform. Decis. Mak.*, 19, 57 (2019). DOI: 10.1186/s12911-019-0763-6. See also Reinforcement Learning for Clinical Decision Support in Critical Care: Comprehensive Review (see Note 73).

[89] *The Alignment Problem* (see Note 85), p. 39.

A further risk specific to machine learning is that its reliance on hidden correlations is not readily subject to human oversight. Consider the example of using a dataset of hospitalized pneumonia patients to predict survival (see Note 86). That example comes from a case where a developer was building a rule-based AI model, without machine learning. The dataset showed a pattern of longer survival time for pneumonia patients with asthma. This correlation seemed absurd on its face, and it was, because asthma is a major risk factor in pneumonia cases. The developer questioned the doctors on his team, and they explained that the correlation simply reflects that pneumonia patients with asthma are hospitalized quickly (usually going to the ICU), which results in more protective care, which results in longer survival time, which means that asthma positively correlates with pneumonia survival. This correlation is perverse for purposes of an algorithm determining whether pneumonia patients with asthma should go to outpatient or inpatient care. Yet, machine learning could find and use such spurious correlations without that ever being noticed.

It turns out that machine learning models are generally more powerful but less transparent than rule-based models.[90] This has led to efforts to develop machine learning models with greater transparency. One of these models, when later applied to the pneumonia dataset from the above case, showed additional spurious correlations between survival and factors such as chest pain, heart disease, and being over age 100—each of which led to more intensity of care and thus increased survival. As a result, the machine learning model that learned these correlations was never put into use.

The above story about the asthma-survival correlation began 20 years ago. The software developer who explored the hidden correlations in his machine learning model now says: "Everyone is committing these mistakes, just like I have committed them for decades, and didn't know I was doing it. My goal right now is to scare people. To terrify them."[91] He is not alone in such views. "As machine learning models proliferate throughout the decision-making infrastructure of the world, Brian Christian writes, "many are finding themselves uncomfortable with how little they know about what's actually going on inside those models."[92]

Now we turn from supervised and unsupervised learning to RL. It has three key differences[93] from supervised and unsupervised learning.

- Sequential decisions are connected, in that each decision sets the context for the next one. Examples are the decisions made in going through a maze, or playing chess (p. 131). In contrast to RL, decisions are independent of each other in supervised and unsupervised learning.

---

[90] *The Alignment Problem* (see Note 85), p. 84. ("It's often observed in the field that the most powerful models are on the whole the least intelligible, and the most intelligible are among the least accurate.")
[91] *The Alignment Problem* (see Note 85), p. 86.
[92] Ibid.
[93] Ibid., pp. 131–132.

- Reward and penalty information is minimal. The system learns whether or not the ultimate outcome is favorable but does not learn what would have been the optimal choice at each sequential decision point. RL has thus been compared to learning from a critic rather than a teacher.

- Feedback from the ultimate outcome is delayed until the end. It is thus not possible to assign credit or blame for the ultimate outcome to any particular decision along the way.

The review article cited in Note 73 describes some issues and challenges of RL in the context of critical care in ICUs. The article makes clear that designing RL tools is full of uncertainties—that no one should accept the output uncritically. Some of the uncertainties are the following.

- The effect of actions taken is determined by comparing the patient's physiologic state at different times. That state is measured by various features (e.g., demographics, lab values, vital signs) or certain encoding methods. This determination of the patient's state along with the reward function determines what constitutes positive and negative reinforcement, so how the determination is made is crucial.

- Reward functions are typically designed on the basis of existing clinical practice and guidelines, which may be variable and inconsistent (see Section 3.3.1). Moreover, designing a reward function to adequately capture existing practices and guidelines is difficult.

- Estimated mortality is a common metric for evaluating the trajectory of sequential decisions. "However, the problem with the estimated mortality is that it is calculated from simulated trajectories with observational data, and may not be the actual mortality data."[94] Moreover, mortality may or may not be the right evaluation metric, depending on the context. As an alternative to mortality data, "observational data may not truly reflect the underlying condition of patients."[95]

- There are a variety of RL models, not all of which involve deep learning from deep neural networks. Models using deep neural networks have more representational power, to better learn optimal treatment recommendations. In designing an RL tool, it is important to consider the pros and cons of different models in different contexts.

- "RL models trained on a single data set, regardless of the data volume, cannot be confidently applied to another data set," because biases are inherent in a data set obtained

---

[94] Reinforcement Learning for Clinical Decision Support in Critical Care: Comprehensive Review (see Note 73 above), PDF p. 10.
[95] Ibid.

from any one institution. The biases are "due to multiple factors, including operation strategy, hospital protocol, instrument difference, and patient preference."[96]

- Inverse RL assumes that the dataset adequately captures what the expert clinicians have done. That assumption may not be correct, which could compromise the reward function derived from the dataset.

These issues and others indicate that the results of RL should be carefully assessed. A process of continuous scrutiny and improvement for each problem situation is necessary.

3.4   DYSFUNCTIONAL MEDICAL RECORD SYSTEMS

3.4.1   LACK OF RECORDKEEPING STANDARDS FOR MANAGING COMPLEXITY

Strangely, medicine has almost no generally accepted standards of care for the structure and contents of medical records. Instead, recordkeeping depends largely on idiosyncratic choices made in the minds of doctors:

> …much of the content of the record *depends on what the clinician chooses to include*, and thus there may be variations in the extent to which clinical reasoning is documented (e.g., what alternative diagnoses were considered, the rationale for ordering [or not ordering] certain tests, and the way in which the information was collected and integrated).[97] [Emphasis added.]

It's true that "there are some common conventions for structuring medical records (both in paper and electronic formats)," as the cited report observes. But those conventions are not enforceable standards of care based on clinical quality. Further variability results from "regulatory and local rules affect[ing] which members of the diagnostic team contribute to the documentation in a medical record and how they contribute," *ibid*. Those rules vary by jurisdiction. Variability of that kind is clinically random.

This was the situation when LLW entered medical school in the mid-1940s, and later when he conducted research in biochemistry. Contrasting his experiences in scientific research and medical practice, he found that doctors learned a core of medical knowledge but not a core of scientific behavior. For example, scientists keep meticulous records of their data, they focus on defined issues, they systematically formulate and test hypotheses under controlled conditions, they do not accept arbitrary time limits on their investigations, and their publications follow established formats. Such disciplined practices are largely missing from medical practice.

---

[96] Ibid., PDF p. 11.
[97] Improving Diagnosis in Health Care (see Note 19), p. 102.

Seeing the need to bring scientific rigor to medical practice, LLW concluded that the medical record was key. The record should be not just a repository of patient data but an information tool for assembling structured information and guiding the decision process.

This situation persists more than 50 years after LLW introduced recordkeeping standards, referred to here as the problem-oriented health record (POHR) rather than the usual label, problem-oriented medical record (POMR).[98] The substance of LLW's standards is discussed in Chapter 9. Here we further explain the clinical needs served by medical records and how those needs go unserved by traditional and current medical records.

For some medical problems, assembling the right information quickly reveals a solution. See Chapter 8. For complex problems, however, the right solution is not immediately apparent. These are the cases where the medical record is most vital. Multiple diagnostic or treatment options must be investigated, often requiring trial and error with careful monitoring over time. A complex problem often involves several medical specialties. The difficulty of handling a single complex problem escalates when patients have multiple problems, as typically happens. Indeed, multiple problems are characteristic of high-cost populations—older patients and those with chronic illness. Their problems, and the medical interventions for each, frequently interact. The interactions can easily derail what might otherwise be well-conceived plans for each problem considered in isolation. So no single doctor has the expertise to handle this kind of complexity.

Another difficulty in complex cases is the gap between medical knowledge and individual patients. Medical knowledge is expressed in generalizations about large populations. These at best only approximate, and often distort or hide, the uniqueness of individual patients. The result is a dangerous gap between what doctors are educated to expect (their "knowledge") and the realities they actually encounter. Bridging this gap is central to the care of complex patients. Somehow, care must be managed in a way that comprehends both general knowledge and patient individuality. Yet, the human mind's natural tendency is to comprehend the former more than the latter. Overcoming the mind's tendency to generalize requires external tools. (Indeed, those tools are needed to comprehend the generalizations of medical knowledge, which are themselves too complex for the mind to manage.)

Beyond these analytical difficulties, complex cases present two logistical difficulties. First, numerous symptomatic and physiological variables, plus care plans and medical actions, must be tracked over time, usually over a period of years when chronic illness is involved. "The volume of data on a chronic patient becomes so large that it becomes unmanageable, and therefore lost, just

---

[98] The POHR label is consistent with current usage of the terms electronic medical record (EMR) and electronic health record (EHR), the latter being broader in scope than traditional medical records. See this blog post, EMR vs. EHR – What is the Difference? This 2011 post was written before the recent focus on SDoH. Now that SDoH data are increasingly being recorded and used, the EMR-EHR distinction is especially relevant.

as the volume of data in the medical literature is already unmanageable and lost to the average practitioner," as Dr. Ken Bartholomew has observed.[99]

The second logistical difficulty in complex cases is the need to coordinate care among multiple practitioners at multiple sites over time, while communicating with the patient (and/or the family) throughout. Patient communication is fundamental, especially in chronic illness. Without it, patient awareness, participation and commitment are undermined. Unavoidable complexity must somehow be made manageable by patients who need to cope with what is happening to their own bodies and minds.

For this, patients need medical record tools that they and their providers can jointly use for guidance. This need is greatest with chronic disease. As LLW observed 50 years ago, long before "patient empowerment" was commonly discussed: "In the last analysis, the patient with a chronic disease must in large part be his own physician … patients are the largest untapped resource in medical care today."[100]

3.4.2   TWO EXAMPLES OF WHY BETTER EHRS ARE ESSENTIAL AS TOOLS FOR PROCESS GUIDANCE AND COMMUNICATION

Consider the following two examples of how lack of sound medical recordkeeping can hinder provider communications and informed patient/family involvement. What was missing in both of these cases was not informational guidance (the clinicians involved had the requisite medical knowledge). Instead, what was missing was a system for process guidance, specifically, a recordkeeping system to communicate a critical task for action.

3.4.2.1 An Infant's Death from Cancer

The first example is the subject of a lawsuit filed in July 2020.[101] After Francesca Webster gave birth prematurely to twin daughters Emma and Zoe, during care at the hospital's neonatal intensive care unit, ultrasounds showed a cystic mass in Emma's right adrenal gland. A nurse practitioner wrote two notes in the medical record, one stating that the mass could be neuroblastoma (a cancer), and the other recommending a follow-up ultrasound. But that follow-up failed to happen. The record contains no indication that anyone ever communicated the information to Ms. Webster or her husband, and they say they were never informed. (They should have been informed upon discharge from the hospital; the discharge documents should have recorded that communication and been

[99]   Bartholomew, K., The Perspective of a Practitioner, in Weed, L. L. et al., *Knowledge Coupling: New Premises and New Tools for Medical Care and Education* (New York: Springer-Verlag, 1991), p. 273.

[100]   Weed, L., *Medical Records, Medical Education and Patient Care* (Case Western Reserve University Press, 1969), pp. 46, 48.

[101]   Maryland family files lawsuit against Johns Hopkins Hospital alleging negligence led to 2-year-old twin's death, *Baltimore Sun*, July 24, 2020.

provided to the parents.) As a result, the cystic mass went unexamined and untreated for nearly two years. The aftermath, as described by the *Baltimore Sun*, was predictable.

> By July 2018, Emma began experiencing fevers, stomach problems and congestion. Abdominal imaging conducted at the end of that month showed liver and spleen enlargement and displacement, according to the complaint. An ultrasound at Hopkins showed a large abdominal mass, noted to be suspicious for neuroblastoma.

> Further investigation found Emma had cancer throughout her body, including in her abdomen, lymph nodes, upper chest, neck, skull, pelvis, spine and left femur, which originated at the site of her adrenal cyst. After undergoing chemotherapy and radiation, Emma died about 10 months later, on June 3, 2019.

The parents sued, supported by an expert's affidavit that prompt follow-up would have led to early detection of the cancer, surgery to remove it, and likely survival.

What this case shows is not just errors by providers. Primarily it shows the absence of a system to protect against the inevitability of error. No one would argue that parents like Mr. and Mrs. Webster should be expected to anticipate mistakes, take the initiative to obtain the medical records, and scour them for important information that might not have been communicated. Similarly, no one should expect every individual provider to identify every action that might be needed and then follow up to make sure that someone took every needed action. Instead, there must be a system of care assuring such reliability.

The core infrastructure for that system of care must be the medical record. What that means is the subject of Chapter 9.

3.4.2.2 A Father's Near-Death from Sepsis

The second example[102] involves the father of two distinguished doctors, Dr. Robert Pearl, a former CEO of the country's largest medical group, and his brother Ron, chair of the anesthesia department at Stanford Medical School. They found themselves watching over the care of their father in the ICU at Stanford Medical Center. He was unconscious, in septic shock from an unidentified infection throughout his body. Then blood tests revealed a pneumococcal infection. This was mystifying, because their father had been vaccinated against pneumococcal infection years earlier, after surgery to remove his spleen. Or so everyone thought. Dr. Pearl writes (p. 7):

> But after calling around, I discovered the hard truth. My dad's doctors in New York assumed the ones in Florida had given him the vaccination. The physicians in Florida assumed the ones in New York had done so. The medical specialists believed the surgeon who removed my father's spleen had administered it. And all the specialty

---

[102]   From Robert Pearl's book *Mistreated* (see Note 26), pp. 1–9.

physicians thought my father's internal medicine doctor had taken care of it. In the end, no one had.

Due to this oversight, Dr. Pearl's father almost died. He spent four days in the ICU and ten more days in the hospital. He survived but was permanently impaired. Before this incident, Dr. Pearl writes:

> …my father was the most energetic person I know. But as he takes his first steps outside the medical center, it's as though he has aged an entire decade. … the toll this experience has taken on his body will sap our father's strength for the rest of his life. We know he will never again be the man he was before [p. 8].

Dr. Pearl emphasizes that all the doctors involved were fully aware of the need for vaccination after spleen removal (although such common awareness is sometimes lacking). But the vaccination task fell between the cracks. That failure was the outcome of two deficiencies: (a) lack of a health record tool to inform everyone of what needed to be done; and (b) lack of procedures for assigning responsibility, scrutinizing what is done, and following through on any oversights, all of which the tool would greatly facilitate. Chapter 9 discusses the specifics of the missing health record tool and how it could be used to help establish and enforce the missing procedures. Because the tool would be used for all recordkeeping, the procedures would have uniformity and simplicity.

Ideally, Dr. Pearl's father would have had a single EHR, used jointly by all providers and him (including family members or other patient representative). That EHR record should have included a plan for post-surgery follow-up care, including the vaccination task. That task should have been automatically proposed by the EHR (or a connected tool) for entry in the record. Such automatic guidance was feasible in advance because the need for pneumococcal vaccination after spleen removal was well-established knowledge. Had the EHR tool automatically proposed that guidance, then it could have been reviewed and sent to the clinicians involved and the patient as a task to be performed and entered in the record. That messaging and recording would facilitate assignment of responsibility for the vaccination task as a standard operating procedure. Most importantly, the patient/family would be fully informed, and thus empowered to follow up if everyone else dropped the ball.

The system of care just described ideally would involve a single record for the patient used by all providers regardless of their institution. A single record is a large departure from the status quo of scattered records (see the next section) and is not going to be widespread in the immediate future. In the meantime, separate provider record systems should be designed to generate the necessary guidance and messaging, with or without interoperable information exchange. The key is effective guidance, decision making, and action at both ends of that exchange. Moreover, the patient should be included in that exchange, via messaging and access to the record in language the patient can understand. Effective communication between the patient and clinicians is a form of interoperability no less important than the electronic exchanges between different computer systems.

3.4.3   THE "SCATTERED MODEL" OF FRAGMENTED RECORDS

Beyond the internal failings discussed above, health records are scattered among a patient's various providers, both inpatient and outpatient. Fragmenting a patient's records in this way creates a host of intractable problems. Many have hoped that these problems can be solved with interoperability among providers' different computer systems. But that hope is not realistic.[103]

The ultimate solution is to replace the scattered model with a "one-patient-one-record" model. Steps in that direction were attempted by Google, Microsoft, and others more than a decade ago, when they offered repositories for consumers to collect and combine their records into a personal health record (PHR). The PHR approach did not succeed, in large part because it imposed too high a burden on patients to gather their data. Now the time may be ripe for a new version of the PHR approach. We discuss this potential solution in Section 9.2.

3.4.4   CLINICIAN DISILLUSIONMENT

For years clinicians have been disillusioned with EHRs. EHRs have become so cumbersome and clinically dysfunctional that they inflame clinician burnout. Burnout was already a serious issue before EHRs. National Physician Suicide Awareness Day is evidence of how serious the burnout issue has become. Nurses as well as doctors have long faced this issue and continue to face it with EHRs.[104]

The toxic effects of EHRs on clinicians has been a dirty secret in the health care industry. But now the situation is increasingly being exposed to the general public. In November 2018, *The New Yorker* published "Why Doctors Hate Their Computers," by Atul Gawande, the best-selling doctor-author. In March 2019, *Kaiser Health News* and *Fortune* magazine jointly published "Death by 1,000 Clicks: Where Electronic Health Records Went Wrong," by the journalists Fred Schulte and Erika Fry.[105] In March 2020, *The New Yorker* published Dr. Siddhartha Mukherjee's article, "What the Coronavirus Reveals About American Medicine" (see Note 2), which discussed the failings of EHRs, among other topics.

---

[103]   For a detailed critique of the scattered model, see Challenges of the TEFCA Scatter Model show the Vital Need to Establish the Individual at the Center, and Comments on: Trusted Exchange Framework and Common Agreement (TEFCA), Draft 2, both dated June 17, 2019 and both authored by Gary Dickinson, Co-Chair, Health Level Seven (HL7) EHR Work Group. The draft TEFCA guidance is available here.

[104]   Jason, C., Nurses need a more significant say in regards to EHR burnout. *EHR Intelligence* (January 13, 2021).

[105]   See also "Botched Operation" for related videos accompanying the *KHN/Fortune* article's full text, and Schulte's summary of five takeaways from the article.

In different ways, these articles are impressive—but also seriously incomplete. They are written as if the failings of EHRs are a matter of recent history. This view misses an earlier and deeper reality.

The recent history is that the 2009 "HITECH Act" (part of the stimulus legislation enacted during the financial crisis) paid providers tens of billions of dollars to upgrade their medical record systems to certified EHRs, conditioned on complying with new federal regulations for "meaningful use" of EHRs. But the earlier reality was that both paper and electronic medical records were always clinically dysfunctional, and thus vulnerable to compromise by financial concerns.[106] And the meaningful use regulations failed to fix that.

Dysfunctional medical records were rooted in lack of standards and discipline for managing clinical information. LLW realized that poor medical records were both a cause and effect of this disorder. Comparing this situation to his experience in biochemistry research, where recordkeeping integrity is paramount, LLW concluded that better medical records were key to effective medical decision making.

So, in the mid-1950s, LLW began to conceive standards to govern design and use of health records. His concepts evolved to include:

- clinical standards of care defining the required elements of the record, the required contents of those elements, and the required structure for organizing those elements;

- electronic tools designed to implement the clinical record standards in patient care; and

- governance mechanisms for enforcing the standards and use of the corresponding tools, together with corresponding reforms in medical education, licensure, and regulation.

LLW's concepts were largely ignored in the meaningful use regulations issued under the HITECH Act. So the regulations failed to define precisely what is needed from EHRs.

Also missing are standards to guide informational inputs to medical records. Without those standards, each doctor's personal judgment continues to determine the contents of the record, whether paper or electronic. That is, each doctor judges what to ask the patient about, what to look for in physical exams, what lab tests to order, what diagnoses to make, what treatments are needed, what to enter in the record, and how to enter it. These judgments have become somewhat regulated by requirements imposed for insurance reimbursement purposes, especially under Medicare. But those requirements fall far short of what is needed for clinical purposes (the subject of Chapter 8).

The outcome is a state of confounding complexity and confusion, a Tower of Babel. That outcome is what *The New Yorker* and *Fortune* articles portray.

---

[106]   See *Medicine in Denial* (Note 4), pp. 37–38.

3.5 MEDICAL KNOWLEDGE AS COMPROMISED BY THE HUMAN MIND

All the cases in this Chapter 3 illustrate that the human mind's failings are built into medical knowledge—itself an element of the health care system. The development of medical knowledge does not just involve advances in medical science. It also involves how social systems for medical education, licensure, and regulation of practitioners coped with the escalating informational burdens on doctors. Those burdens became increasingly untenable (and doctor training increasingly prolonged and expensive) with the rapid advances in medical science during the 20th century. In response, medical practice became increasingly specialized. This took the form of categories of expertise defined by body systems (cardiovascular, musculoskeletal, endocrine, etc.).

But specialization in this form does not fit with patient needs. Body systems are highly interconnected. Patient problems therefore cross specialty boundaries. Doctors who stay in their own specialty lanes tend to develop tunnel vision. The "care" they provide is not patient-centered, nor is it "problem-oriented" as that concept is presented in Chapter 7.

In a problem-oriented system, medical knowledge should become more and more individualized, reflecting the uniqueness of each person's physiology, psyche, and circumstances. But in the current non-system, knowledge tends to be reduced to population-based generalizations. This reductionism happens in part for short-sighted economic reasons (for example, practice guidelines treat an inexpensive drug for a population as more cost-effective than enabling each person's choice of drugs to be optimized for individual needs).[107]

But the deeper reason for this reductionism is that the human mind can more easily cope with a few generalizations about a population than with countless differences among individuals. This reductionism distorts scientific knowledge. Each person is a unique combination of resemblances to and differences from other people. Medical science explores hidden resemblances, understanding of which improves diagnosis and treatment. But effective problem solving very often requires understanding individual differences as well as resemblances. We therefore need a system where health decision making and knowledge development both become increasingly individualized, not standardized. That approach is in the long run the most cost-effective, if long-term effectiveness as well as short-term costs are taken into account.

Reinforcing these points is the fact that "current models for health-care delivery and quality measurement woefully underestimate the complexity of individual patient variation," as Dr. James

---

[107] This example just scratches the surface of the myriad ways economic and cultural forces may compromise medical knowledge. For example, drug companies may promote a supposed "disease" in order to market a drug that mitigates that disease. Providers may engage in "upcoding" to maximize reimbursement. Or medical specialty societies may influence approval of lucrative procedures for the same purpose. Such practices in turn compromise the integrity of medical records, problem lists in particular, where real clinical problems are skewed or obscured by reimbursement rates and policies.

Sorace has written.[108] He cites a study of the distribution of multiple comorbidities in the Medicare population and another study of identical twins in that population. These studies showed "that there is no average or typical Medicare patient." Patients have countless combinations of co-morbidities, in part because rare diseases are collectively common. Thus, clusters of patients with the same co-morbidity patterns are large in the number of clusters and small in the number of patients within each cluster. For patients with rare diseases, the average cluster size is only five patients, "and these patient clusters account for nearly 80% of expenditures." Moreover, the twin study shows that "heredity plays a surprisingly small role in the chronic diseases found in the Medicare population. Patients, including family members, are very individual in their pattern of comorbidities." These complexities, Dr. Sorace writes:

> …greatly complicate the development of useful practice guidelines and clinical [quality] measures for complex patient populations, and this explains the challenges in measure development discussed above. Significantly, the studies discussed above concluded that less common diseases, in aggregate, are important drivers of comorbidities and expenditures, and explain the challenges physicians face when caring for specific patients. Even the most experienced physicians are constantly seeing new presentations and combinations of comorbidities. … Not even the largest health-care networks have adequate data to attempt to optimize individual care for many patients …

This heterogeneity defeats practice guidelines and protocols derived from clinical trials on randomized populations (which exclude much individual variation). Dr. Sorace thus advocates "a decentralized crowdsourced environment that enables the collaborative care of the numerous small clusters of patients." This approach "must also prioritize the accurate diagnosis of individuals" in order to identify the clusters of patients with similar sets of co-comorbidities. "Unfortunately, diagnostic errors are far too common in our health-care system to effectively support the treatment of rare patient clusters. … it will become increasingly important to reward *technologies that improve the scientific basis of diagnosis.*" [Emphasis added.]

The CDS and EHR tools and standards described in Chapters 7–9 would go far toward achieving the goals stated by Dr. Sorace.

---

[108] Sorace, J., Payment reform in the era of advanced diagnostics, artificial intelligence and machine learning. *Pathol. Inform.*, 2020;11:6. DOI: 10.4103/jpi.jpi_63_19.

CHAPTER 4

Magnitude of the Problem

4.1   QUALITY FAILURES—MEDICINE'S CHRONIC CONDITION

The incidents and issues discussed in the preceding chapter are only the tip of an iceberg of quality failures. And those failures are not a remote threat in an ocean of high-quality care. Instead, the failures are broad and deep—all-encompassing at the macro level and excruciatingly intricate at the micro level. Their consequences for patients range from catastrophic (death or lifetime suffering and disability) to nothing (harmless near misses). The near misses lessen the total harm, but not to a tolerable level. On the contrary, medical errors are reportedly the third leading cause of death in the U.S.,[109] with those deaths being only a fraction of the harm inflicted and those errors only a fraction of misguided and wasteful care.

Consider the subtitle of Dr. Robert Pearl's book, *Mistreated: Why We Think We're Getting Good Health Care—and Why We're Usually Wrong* (discussed in Section 2.2 at Note 26). The subtitle asserts that we are "usually" not getting good care. This is extraordinary. It implies that poor quality care occurs most of the time, and only near misses save us from being *harmed* most of the time (while nothing saves us from endemic waste). In some economic contexts, quality failure rates of a few tenths of a percent are unacceptable. But in medicine, exponentially greater rates of quality failure and waste are "usual," as Dr. Pearl says. This is consistent with feedback he receives:

> Whenever I speak at health care conferences about my dad's experience [see Section 3.4.2.2], people nod knowingly. I'm always amazed by how many from the audience will line up afterward to share similar stories and frustrations about

---

[109]   Makary, M. and Daniel, M., Medical error—the third leading cause of death in the U.S. *BMJ* 2016;353:i2139 DOI: 10.1136/bmj.i2139. These analyses do not quantify what portion of this toll is attributable to decision making (the subject of this book). Execution of decisions is equally important. *To Err Is Human* (Note 3) states (pp. 54–55):

> For this report, *error is defined as the failure of a planned action to be completed as intended (e.g., error of execution) or the use of a wrong plan to achieve an aim (e.g., error of planning)*. From the patient's perspective, not only should a medical intervention proceed properly and safely, it should be the correct intervention for the particular condition. This report addresses primarily the first concern, errors of execution, since they have their own epidemiology, causes, and remedies that are different from errors in planning. Subsequent reports from the Quality of Health Care in America project will consider the full range of quality-related issues, sometimes classified as overuse, underuse and misuse.") [Emphasis in original].

The next IOM/NAM report, *Crossing the Quality Chasm* (National Academy Press, 2001), discussed medical decision making at length, with multiple references to LLW's work.

the care their loved ones received. Just about every family seems to have had an experience like ours.[110]

The enormity of quality failures reflects the inherent and increasing complexity of medicine. Health professionals individually are doing all they can to cope. They are "socialized to strive for error-free practice," and are "probably among the most careful professionals in our society."[111]

But their prodigious efforts are insufficient. Medicine's high volume of high risk activities makes significant levels of failure and harm seem inevitable. To take but one example cited in Dr. Leape's article, ICU patients "were the recipients of 178 'activities' per day," with "an average of 1.7 errors per day per patient, of which 29% had the potential for serious or fatal injury." In that light, the error rate—about 1%—may not seem high. No doubt it would be considerably higher but for the dedication of professionals working in ICUs. But, Dr. Leape writes:

> …a 1% failure rate is substantially higher than is tolerated in industry, particularly in hazardous fields such as aviation and nuclear power. As W.E. Deming points out …, even 99.9% may not be good enough: "If we had to live with 99.9% we would have: 2 unsafe plane landings per day at O'Hare, 16,000 pieces of lost mail every hour, 32,000 bank checks deducted from the wrong banks accounts every hour.

From this perspective, Dr. Leape's 1994 article made a now-famous comparison: reported rates of fatal injuries from iatrogenic care are "the equivalent of three jumbo-jet crashes every two days."[112]

Much of this toll results from errors of execution rather than decision making. But the magnitude of decision-making failures is still huge. Diagnostic error alone is reportedly so common that "likely … most people will experience at least one diagnostic error in their lifetime, sometimes with devastating consequences."[113] And so "in large studies of outpatient malpractice claims, diagnostic errors emerge as the most common category."[114] According to the Society to Improve Diagnosis in Medicine,

> All told, diagnostic errors affect an estimated 12 million Americans each year, and likely cause more harm to patients than all other medical errors combined. Missed diagnoses also lead to higher healthcare costs—through treatment of sicker patients with more advanced disease; by overuse of unnecessary, expensive diagnostic tests; as a consequence of malpractice claims; and with the high costs of treatments for diseases patients do not actually have. Some have estimated that $100 billion or more may be wasted annually in the U.S. as a result of inaccurate diagnosis. … An estimated 40,000 to 80,000 people die each year from diagnostic failures in U.S.

---

[110] *Mistreated*, (see Note 26) p. 9.

[111] Error in Medicine (see Note 32), n. 17, citing a 1989 unpublished study.

[112] Error in Medicine (see Note 32).

[113] Improving Diagnosis in Health Care (Note 19), p. 355.

[114] Singh, H., Schiff, G. D., Graber, M. L., et al., The global burden of diagnostic errors in primary care. *BMJ Qual. Saf.*, 2017;26:484–494. See also the sources cited by *Medicine in Denial*, p. 19 n.6.

hospitals alone, and probably at least that many suffer permanent disability. The total across all clinical settings is likely much higher.[115]

As to the nature of decision making errors, a 2013 study of malpractice cases explains:

> Most failures were in diagnosis (mostly cancer) and can be pinpointed to just 3 broad areas of outpatient activity: (1) *obtaining and regularly updating a comprehensive patient and family history, physical examination, and evaluation of symptoms*; (2) *ordering (or failure to order) diagnostic or laboratory tests*; and (3) *managing referrals and following up with the patient*. It would appear that *a finite number of safeguards could be put in place* to prevent and/or mitigate these errors, which, in turn, could have a major effect on patient safety and malpractice. Testing such measures as (1) *structured methods* for reliably obtaining and updating patient or family history (including patient self-completed data) and physical examination records, (2) use of *differential diagnosis reminders and checklists*, and (3) implementation of *more reliable test ordering and follow-up strategies*, along with methods for ensuring specialty referral and follow-up and improved documentation in these 3 realms, offers avenues to address shortcomings found in these cases. *Optimally designing electronic medical records to more reliably and efficiently facilitate reengineering these functions and testing their effectiveness represents a priority for preventing future malpractice.*[116]

These are the kinds of issues directly addressed by the tools and standards in Chapters 7–9.

Before modern medicine, high risks of disease, injury, and premature death seemed inherent in the human condition. Enormous progress against these risks has been made, but modern public health (not medical practice) accounts for the bulk of this progress. Now these high risks of disease, injury, and premature death seem inherent in advanced medical practice and cutting edge medical science, with all their metastasizing complexity. Just as lack of pandemic preparedness made us vulnerable to COVID-19, so lack of preparedness to manage complexity makes us vulnerable to the risks arising from modern medicine.

The problem's magnitude goes beyond the medical harm inflicted. *Economic* harm from wasteful, unnecessary, ineffective, and other low-value care is rooted in *medical* quality failures—specifically, defective decision making, either by providers who choose that care or else by consumers who choose it under the influence of providers and third parties. Both providers and consumers lack a system of care equipping them with the tools and standards they need for their decisions. Medical and economic harm are also rooted in defective execution of decisions by providers, which in turn results from lack of a system to assure decisions are executed reliably. In short, failures of quality and economy in medicine are different sides of the same coin.

[115] https://www.improvediagnosis.org/facts/.

[116] Schiff, G. D., Puopolo, A. L., Huben-Kearney, A., et al., Primary care closed claims experience of Massachusetts malpractice insurers. *JAMA Intern. Med.*, 2013;173(22):2063–2068. DOI: 10.1001/jamainternmed.2013.11070.

All this might suggest that failures of both quality and economy in medicine are a truly in-tractable problem.[117] The most we can hope for, it seems, is tackling failures one-by-one, triaging as best we can so that our efforts have the most effect—almost like medics in battlefield and disaster situations.

But such pessimism has never taken over. Modern medical culture, to its credit, has never given up systematic efforts to improve quality.[118] In fact, as LLW witnessed over his long career, every generation of clinicians rediscovers the magnitude of quality failures and embarks on new crusades against it. A mid-20th century example of rediscovering the issue is described by Michael Millenson, writing on obstetrical practice during the 1950s.

> Today many doctors speak about a "golden age" of medicine that supposedly existed in the pristine fee-for-service days before Medicare, Medicaid, utilization review, and managed care. The theory is that physician autonomy allowed doctors to con-centrate on caring for patients instead of worrying about pettifogging paper require-ments. That is akin to believing that the 1950s were a golden era for pharmaceuticals because the Food and Drug Administration had not yet required tedious clinical trials to demonstrate that a drug was effective. …
>
> One of the most dramatic examples presented a horrifying picture of casualness about a procedure that takes away a woman's ability to bear children. In reviewing the records of more than 6,000 hysterectomies in thirty-five hospitals, physician James Doyle found hundreds of women who had received either no preoperative diagnosis or a simple diagnosis such as "pain." Postoperatively it turned out that 30% of all of the patients ages twenty to twenty-nine who were subjected to hysterectomy had no disease whatsoever, a number that Doyle rightly called "appalling."[119]

The patient safety movement arising from Lucian Leape's 1994 "Error in Medicine" article and the Betsy Lehman case was a later generation's re-discovery of quality failures in medicine. "The

[117]   See Mafi, J. N. and Parchman, M., Low-value care: an intractable global problem with no quick fix. *BMJ Qual. Saf.*, 2018;27:333–336.

[118]   For a history of quality improvement efforts in American medicine, see Millenson, M., *Demand-ing Medical Excellence* (University of Chicago Press, 1997). See also Millenson's 2013 follow-up blog post, Still Demanding Medical Excellence, where he pointed to various signs of progress but also wrote: "the evidence is irrefutable. Tens of thousands of patients have died or been injured year after year *because readily available information was not used*—and is not being used today – to guide their care. … even today, many in the quality field hesitate to stare directly into the face of the grim statistics on the persistence of the medical error problem. … Most recently, a September 2013 study in the *Journal of Patient Safety* put the toll at 210,000–400,000 deaths per year." [Emphasis added.]

[119]   Millenson, M., "Miracle and wonder": the AMA embraces quality measurement. *Health Aff.*, Vol. 16, No. 3 (May/June 1997). DOI: 10.1377/hlthaff.16.3.183. Note 8 cites Doyle, J.C., "Unnecessary hysterectomies: study of 6,248 operations in thirty-five hospitals during 1948," *JAMA* (January 31, 1953): 360–365.

link between the patient safety movement and the older, better-established quality improvement movement has often been uncertain: At times, the two groups have seemed to be entirely unaware of one another."[120]

Although generations of observers have railed against medicine's ongoing quality failures, and generations of clinicians have struggled to overcome them, none have clearly seen what LLW saw half a century ago—that failure is built into the doctor's role. Nor have others conceived, as LLW has, the operational specifics of an alternative to that role.

4.2   ECONOMICS OF THE PROBLEM: THE HIGH COSTS OF LOW QUALITY

Readers sympathetic to the solutions described in Part II below may be concerned that the economic payoff is too uncertain and the change involved too great for the solutions to be worth undertaking. Such a view underestimates the magnitude of the problem and the power of the solution.[121]

The extent of avoidable error and waste in medicine has yet to be fully grasped, for many reasons. Dramatic advances in medical science dominate awareness. In medical practice, a truly effective system of care has yet to be widely experienced, so we don't have a basis for comparing the status quo. Dedicated and talented health care workers apply medical knowledge as best they can and succeed more often than not. Flawed tools and premises block full understanding and awareness of the nature and extent of failure. The worst failures thus seem to be isolated events, the residue of human fallibility in the face of enormous complexity, an inevitable reality that we have no choice but to tolerate.

Yet, tolerating even isolated errors can increase costs exponentially. This occurs at multiple levels.

- When an inexpensive medical procedure is also ineffective, its low cost represents economic waste, not savings.

- Ineffective medical procedures are not only wasteful in themselves but typically escalate costs, because they set in motion further trial and error.

- That trial and error is not only expensive itself but may harm the patient, thus triggering further medical interventions with additional cost and risk. Mishaps may occur every step of the way.

---

[120]   Bates and Gawande, Note 14.
[121]   Much of the following discussion is adapted from Weed, L. L. and Weed, L., Opening the black box of clinical judgment, Part IV: An Economic Perspective (p. 2). *Brit. Med. J.*, *eBMJ* edition, Vol. 319, Issue 7220, November 13, 1999. See also pp. 5–8 of that article for related discussion of Kenneth Arrow's seminal article, "Uncertainty and the Welfare Economics of Medical Care."

- Costs can also escalate when a missed diagnosis or incorrect treatment delays an effective alternative, allowing the patient's condition to worsen.

- The economic ripple effects of all this are incalculable. Resources are diverted to ineffective medical care from more productive uses, patients are delayed from returning to productive activity, and family members are diverted from their own productive activities. Poor use of existing medical knowledge generates unnecessary payment disputes and malpractice litigation, which dissipate resources and lead to more decisions of poor quality (e.g., "defensive medicine" by providers, misplaced reliance on "alternative medicine" by patients). Missed diagnoses and ineffective or harmful treatment decisions can be emotionally harmful to patients, who may be wrongfully branded with psychological diagnoses. Such outcomes disrupt family and work relationships, further reducing the economic value of medical "care." All this generates distrust of providers and the health care system, which deters people from seeking the care they need.

In short, the cost of failure may far exceed any immediate savings from tolerating failure. And tolerance of failure on a large scale is built into everyday medical decision making, with its misguided dependence on the minds of doctors and limited use of essential patient inputs. Failure includes not just outright error but massive amounts of care that are not justifiable in terms of risk, need, cost, benefit, and patient wishes.

"The truth is that U.S. health care is awash in excess, so much so that, to maintain the financial performance they've come accustomed to (and that the market has come to expect), many or most health care organizations now **depend on** egregious unit pricing and unnecessary services" (emphasis in original).[122] Unnecessary services themselves reflect poor quality decision making. Even when services are necessary, their execution may be poor—another area of failure on a large scale.

The persistence of quality failure in disparate economic environments shows the need for fundamentally new tools and approaches. New tools can be powerful in part because they counteract the "cost disease" identified by the economist William Baumol. He argued that that "the amount of physician time spent per patient visit or per illness" cannot decline substantially, and therefore health services "are inherently resistant to automation." On this view, higher costs per unit of health services are required if physicians' relative incomes and the quality of the services they render are to be maintained, making disproportionate increases in health care costs an "ineradicable part of a developed economy."[123] This view is fallacious. As we wrote about Baumol's view in 1994, information resources "need not be subject to scarcity constraints," because information in electronic form can be freely disseminated and made usable without scarce physicians. Baumol wrongly assumed that use

---

[122] When Health Care Organizations Are Fundamentally Dishonest (see Note 52).

[123] Baumol, W., "Anatomy of an illusion. Do health care costs matter?" *New Rep.*, November 22, 1993, pp. 16–18.

of health information depends on the traditional physician role, which is the element "inherently resistant to automation."[124]

It is not possible to isolate and measure the total economic gains that might be achieved (and losses that might be avoided) by reforming medical practice. But various analyses give some sense of the extraordinary potential benefits from improvements in the health domain generally and medical practice in particular. See, for example, a 2015 blog post by Lawrence Summers and Gavin Yarney, "The Astonishing Returns of Investing in Global Health R&D," discussing the Global Health 2035 report. This post understates the benefits by only addressing advances from R&D such as new medicines, vaccines and diagnostics. The post fails to recognize how mainstream medical practice in developed countries, especially the U.S., is a source of avoidable costs and missed opportunities.

Savings in direct costs may be dwarfed by avoiding the indirect costs and opportunity costs of the medical practice status quo. One obvious example of indirect costs is avoidable malpractice litigation. Another example is clinician burnout, which is estimated to cost the health care system $4.6 billion per year. "Substandard EHR usability has consistently been cited as one of the top contributors to clinician burnout." EHR usability can be significantly improved by enabling a problem-oriented view of EHR data (see Section 9.1), which facilitates data retrieval. In turn, "improvement in data retrieval, a crucial component of EHR usability, has the potential to decrease clinician burnout." This happens because a problem-oriented view can "streamline clinical workflows and allow for more efficient and accurate data retrieval while decreasing cognitive load and improving user satisfaction."[125]

A recent analysis is from the McKinsey Global Institute, How prioritizing health is a prescription for U.S. prosperity (October 2020). Focusing on not the costs of health care for payers but the burden of disease for workers and employers, this report finds enormous potential to improve economic growth.

> We estimate that poor health costs the United States about 16% of real GDP annually from premature deaths and lost productive potential among the working age population.

---

[124] Weed, L. L. and Weed, L., Reengineering medicine, *Federat. Bull.: J.Med. Licens. and Discipl.*, 1994; 81:147-83, at p. 150, https://jmr-archives.fsmb.org/Archive/1990s/Journal%20of%20Medical%20Licensure%20and%20 Discipline(Vol81N3).pdf.

[125] Semanik, M., et al., Impact of a problem-oriented view on clinical data retrieval. *J. Am. Med. Inform. Assoc.* February 10, 2021;ocaa332. DOI: 10.1093/jamia/ocaa332. This article is further discussed in Note 206, Section 9.1.2.1. See also Weed, The Problem-Oriented Health Record (POHR) and a Pathway to Reducing Clinician Burden (an August 17, 2020 presentation to the HL7 Reducing Clinical Burden Project, two versions of which are available at https://wiki.hl7.org/Reducing_Clinician_Burden#HL7_RCB_Project_-_Presentations_and_ Reports).

But it does not have to be that way. Rethinking health as an investment, not just a cost, holds the potential not just to improve the health of millions but to accelerate economic growth for decades to come.

The McKinsey report makes clear that a major source of improvement is "through better delivery of *known interventions*" and "deploying *existing approaches* to improve health and prevent and treat diseases" (emphasis added). Specifically, the U.S. "could reduce its disease burden by as much as one-third by 2040" (pp. 2, 4), even before taking into account as yet unknown advances in medical science.[126]

Consistent with the McKinsey report, this book offers a pathway to better delivery of existing approaches already known to the system. That can be achieved only by reducing the system's dependence on what expert clinicians personally know or don't know. The system's informational supply chain must deliver what the *system* knows to patients and clinicians jointly. This can be done through tools designed to apply the system's knowledge to the patient's data. Information tools designed for this purpose must become core infrastructure that patients and clinicians use jointly every day. This whole approach is utterly different than delivery of knowledge through *educating* clinicians in the hope that they will recall and apply it effectively.[127]

If that infrastructure is not built and deployed, then health care spending will continue to go to waste on a large scale. And that waste has several dimensions of destructiveness. Waste sucks up funds otherwise available for the productive spending advocated by the McKinsey report. It means that population health problems fester and grow worse, which in turn increases suffering and lowers economic growth. And it "feeds into the already severe harms caused by growing income inequality," as Professor David Cutler observes. For those reasons he argues that "we ought to pay at least as much attention to ways of improving efficiency as we do to whether and how people should get covered." Improving efficiency is precisely what reforming medical practice would accomplish. And the improvement would happen on both sides of the efficiency equation—spending less and getting

---

[126]   For background, see Five insights from the Global Burden of Disease Study 2019, *Lancet*, October 17, 2020; 396: 1135–59. See also *Everylife Foundation for Rare Diseases, Lewin Group. The National Economic Burden of Rare Disease Study* (February 25, 2021, https://everylifefoundation.org/burden-study/. This study of 379 rare diseases (only about 5% of known rare diseases) finds that for 2019, "their overall economic burden ... is $966 billion, of which 43% are direct medical costs ($418 billion), ... $437 billion are indirect costs associated with productivity losses, and $111 billion are non-medical costs." This study does not determine what portion of these costs would be avoidable by better delivery of existing approaches.

[127]   Corresponding changes in licensure laws could release large potential for upward mobility for health professionals other than doctors (see Note 228 for an example). This alone might have substantial economic benefits, given the large number and wide geographic distribution of those professionals. (There are 3.1 million registered nurses and several million more other medical practitioners, compared to only 752,000 doctors in the U.S. U.S. Bureau of Labor Statistics, Occupational Outlook Handbook, Healthcare Occupations (2019 numbers).) Breaking the doctor monopoly might also unleash a wave of innovations in delivery of care and quality improvement. See Part VIII of *Medicine in Denial* (Note 4).

more for whatever we do spend. Professor Cutler concludes with a deeper point, one that heightens the urgency of reforming the services our health care spending buys:

> The U.S. is being pulled apart as a country, separating into rich and poor. Every dollar that is spent on medical care is one less dollar available for addressing the problems of an unequal society, and one more dollar that is difficult for much of the population to pay. One of the goals for health policy must be to reduce social and economic disparities, not increase them.[128]

[128] Cutler, D., What is the U.S. health spending problem? (February 14, 2018) *Health Aff.*, 2019;37(3). DOI: 10.1377/hlthaff.2017.1626.

CHAPTER 5

Background: Larry Weed

LLW's initial work on medical records grew out his experience doing basic research in biochemistry. As he described it:

> I realized that a research scientist has one problem and can make time the variable and achievement the constant in his work, whereas a physician on a busy ward has multiple problems and limited time, making the behavior of a true scientist almost impossible. As a scientist, you have a very specific project. That's your research. You work on it and work on it, and you finally get it written up. You get it published in a journal. The scientist works under a disciplined system of review and publication of his work. A physician works in a chaotic system of keeping and organizing data and has no systematic review and correction of his daily work.
>
> … Finally I asked myself: suppose you treated a medical student or resident like a graduate student working on a problem to get a Ph.D. You would say to the student, 'Well, what's your problem?' 'Well, Mrs. Jones is one of my patients.' 'Well, what are her problems? Where's the list of her problems? We have to work up each problem. Well, don't you have a record?' The medical record of Mrs. Jones must be a medical student's scientific notebook. With a scientist doing research you can see in his notebooks in the laboratory what he does each day. You can see his data, what came out of the spectrometer and so on, and you can examine it.[129]

LLW found that medical training tolerated low standards of intellectual behavior as compared to the disciplined behaviors demanded in scientific training and research. Recordkeeping behavior in particular is crucial. Medical records were nothing like scientific data or article manuscripts. Recording customary elements such as the "chief complaint" and "progress notes" was largely left to each doctor's discretion, without defined standards and without periodic scrutiny

---

[129] Wright, A. et al., Bringing science to medicine: an interview with Larry Weed, inventor of the problem-oriented medical record. *J. Am. Med. Inform. Assoc.*, November, 2014, 21(6):964–968. DOI: 10.1136/amiajnl-2014-002776. In this interview, LLW describes how he left a promising research career at Yale in 1956 to go to Bangor, Maine, where he first worked out the POHR basic concepts. See also Jacobs, L. Interview with Lawrence Weed, MD – the father of the problem-oriented medical record looks ahead. *Perm J.* 2009; 13(3):84-89. https://www.ncbi.nlm.nih.gov/pmc/articles/PMC2911807/. See also a video of 1971 internal medicine grand rounds LLW did at Emory. https://www.youtube.com/watch?v=qMsPXSMTpFI. Published June 22, 2012. See also Xu, S. and Papier, A., Returning to (electronic) health records that guide and teach. *Am. J. Med.* July 2018; 131(7):723–725. DOI: 10.1016/j.amjmed.2017.12.048. That article finds "an urgent need to reframe the role of electronic health records back to the fundamental principles espoused by Dr. Weed's original vision as a tool to improve physician communication, decision-making, and patient care."

(audit) of what doctors produced. What they produced, even when legible, was often barely intelligible—little more than "'stream-of-consciousness' notes that almost made it impossible to follow the patient's record."[130]

To remedy this state of affairs, LLW developed what became known as the POMR, referred to here as the problem-oriented health record (POHR). The POHR is a standard of care for organizing the record. Its most well-known components are problem lists and SOAP notes.

Two decades after first conceiving the problem-oriented record, LLW concluded in the late 1970s that it was an incomplete solution. Informational inputs to the record are traditionally determined by the doctor's mind. But the mind is unequal to that task. An external tool optimized for the task is essential. That reality led LLW to conceive and develop "knowledge coupling" tools for coupling medical knowledge with patient data. See Section 7.2.

---

[130]   Dr. Charles Burger, quoted in a December 2005 article in *The Economist* on LLW's work.

CHAPTER 6

Background: Medicine and the Domains of Science and Commerce

Medicine is unusual in combining scientific, commercial, and humanitarian activities on a large scale. The humanitarian dimension is paramount. But that dimension demands that medicine be practiced with scientific rigor and economic efficiency. Without those, universal provision of high-quality care becomes unattainable and unaffordable, and medicine inevitably falls short of its humanitarian ideals. Those ideals are violated when clinicians make or execute decisions at less than high standards of quality, and when institutions allow or pay for unnecessary and other substandard care. These shortfalls breed intolerable harm and waste.

In fact, medicine lags centuries behind the times compared to other fields in the domains of science and commerce. When we distinguish between medicine's two basic components—medical practice and medical science—the lag is quite obvious for medical practice. For medical science, the lag is less obvious, but it still exists, notwithstanding the amazing scientific advances during the last century and a half. The lag exists because medical science could have better harvested advances from, and better disseminated advances to, medical practice, were the latter conducted with scientific rigor. As it is, scientific advances and their dissemination have arisen primarily from basic scientific research and development and public health, not from clinical and health services research and development in medical practice.

We have said that medicine is anomalous in its reliance on expert judgment, and we discussed the lack of scientific behaviors and standards in medical practice. In addition, many have observed that medicine lacks the economic discipline and accountability often found in the commercial domain. These points deserve further discussion.

6.1 THE DOMAIN OF SCIENCE: FRANCIS BACON AND KARL POPPER

Science has always faced a wide gap between limited human capacities and the demands of effective research. To bridge that gap, science uses external tools such as measuring instruments, the microscope, the telescope, and the computer. The same is true of researchers in the applied science of medicine. Yet doctors have lagged far behind in habitually using computer tools as external aids to the mind for medical practice.

Just as scientific instruments extend the powers of human sense organs, so the computer extends the powers of the mind. In recent decades, the mind's powers and limits have been the object of study by cognitive psychologists. Although its powers of instinctive judgment are impressive in some contexts, the mind is "a relatively inefficient device for noticing, selecting, categorizing, recording, retaining, retrieving and manipulating information for inferential purposes."[131] Therefore, science, including medical science, has embraced modern electronic information tool.

> The dominant trend in biomedical science and in medical practice, as in every realm of science, is the increasing value and usage of computers. The data so painstakingly extracted in past years are now, through progress in biomedicine, produced in such volumes as to require computers just to record them. The scientist spends more and more time using the computer to record, analyze, compare and display their data to extract knowledge.[132]

This statement begins by equating biomedical science and medical practice. Yet, the examples given are drawn from science, not practice. Here we need to recognize two distinctions. Using the computer to uncover new knowledge for medical science differs from using it to apply existing knowledge for medical practice. And, within medical practice, using the computer as a component of medical devices to enhance the user's physical capabilities differs from using it as an information tool to empower the mind for clinical decision making.

These distinctions suggest that doctors and scientists differ fundamentally in their approach to limited human capacities. Doctors recognize limits in their capacity for observation and data processing, but not in their capacity for applying medical knowledge. Thus, the most advanced, costly and ubiquitous use of computer technology in modern medicine is sophisticated clinical imaging devices. These devices collect detailed data and use sophisticated software to assemble the data into images of internal organs. By comparison, doctors rarely use computer software to assemble patient data and medical knowledge into options and evidence for medical decision making. Instead, doctors rely largely on personal intellect ("clinical judgment") for this pivotal function.

In contrast to medical practice, science has advanced by developing alternatives to unaided judgment. These developments made possible intellectual operations that would otherwise be prohibitively laborious and prone to error.[133]

---

[131]   Grove, W, and Meehl, P., Comparative efficiency of informal (subjective, impressionistic) and formal (mechanical, algorithmic) prediction procedures: The clinical-statistical controversy. *Psychol., Publ. Policy Law* 1996; 2:293-323, p. 316, at https://psycnet.apa.org/record/1997-02834-005.

[132]   NIH Working Group on Biomedical Computing, *The Biomedical Information Science and Technology Initiative.* 1999. Available at https://acd.od.nih.gov/documents/reports/060399_Biomed_Computing_WG_RPT.htm.

[133]   The development of mathematics, for example, was described in these terms by Alfred North Whitehead. Whitehead further stated a broader principle: "Civilization advances by extending the number of important operations which we can perform without thinking about them." See Section 6.2 for further discussion of Whitehead's principle.

Information is the raw material of science. Yet, when information becomes complex, the mind is unreliable and inefficient, as cognitive psychologists have documented.[134] Moreover, normal human behaviors in using the mind lack the rigor that science demands. To overcome these limitations, scientists have developed a variety of practices. These practices include enforcing habitual use of tools and techniques to aid the mind, and simple standards of thoroughness and reliability. This discipline is essential to scientific progress.

> The dazzling achievements of Western post-Galilean science are attributable not to our having any better brains than Aristotle or Aquinas, but to the scientific method of accumulating objective knowledge. A very few strict rules (e.g., don't fake data, avoid parallax in reading a dial) but mostly rough guidelines about observing, sampling, recording, calculating and so forth sufficed to create this amazing social machine for producing valid knowledge. Scientists record observations at the time rather than rely on unaided memory. Precise instruments are substituted for the human eye, ear, nose and fingertips whenever these latter are unreliable. Powerful formalisms (trigonometry, calculus, probability theory, matrix algebra) are used to move from one set of numerical values to another.[135]

These practices bring rigor and reliability to the raw material of science—information. This is achieved by compensating for the limited abilities and variable habits employed in creating, using, recording, and manipulating information. That compensatory function also empowers the mind's creative capacities for judgment and imagination, but its first purpose is to enable trustworthy information processing.

Thus far we have discussed how tools and techniques for aiding the mind bridge the gaps between human cognitive limits and the complexity of science, between normal human behaviors and the rigorous habits of careful investigators. But there are other gaps that science must bridge: gaps between individual, subjective experience and shared, objective knowledge, between limited individual capacities and the greater capacities of cooperative social endeavors.

Remarkably, Francis Bacon envisioned these dimensions of scientific culture at its birth four hundred years ago. As the first thinker who systematically examined the mind's role in the advancement of science, Bacon recognized that external aids to the mind are pivotal: "It is by instruments and other aids that the work gets done, and these are needed as much by the understanding as by the hand." But medicine is in denial of the need for this kind of external aid. So Bacon anticipated the culture of modern medicine 400 years ago when he wrote, "while we wrongly admire and extol the powers of the human mind, we fail to look for true ways of helping it."[136]

---

[134] Grove and Meehl, Note 131.
[135] Grove and Meehl, Note 131.
[136] See Note 13.

Bacon was reacting against academic and ecclesiastical dogma, with its static dependence on the minds of ancient authorities (Aristotle in particular) and its reliance on intellect (through formal, scholastic disputation) as a sterile mode of inquiry. He became deeply skeptical of abstract thought divorced from observation and experience. The learning from experience by those engaged in commercial and practical activities enormously impressed Bacon. He also witnessed a flowering of intellectual life outside the universities. He came to view science and practical learning as cumulative, collaborative activities, anchored in experience, freed from received authority and the individual mind.[137]

Analysis of the mind's limits was central to Bacon's philosophy. Anticipating several currents of 20th century thought (including the cognitive psychology we have already discussed), he identified four "idols of the mind" that distort human thinking and perception.

- **Universal traits of perception that "lie deep in human nature itself."** Therefore, "all our perceptions, both of our sense and of our minds, are reflections of man, not of the universe, and the human understanding is like an uneven mirror that cannot reflect truly the rays from objects …"

- **Individual dispositions and acquired beliefs.** So "each of us has his own private cave or den, which breaks up and falsifies the light of nature."

- **Misuse of language.** This "obstructs the mind to a remarkable extent. … Indeed, words plainly do violence to the understanding, and throw everything into confusion, and lead men into innumerable empty controversies and fictions."

- **"Various dogmas" in philosophy and the sciences.** These "have become established through tradition, credulity and neglect."[138]

Bacon saw a path that led away from the alchemy and astrology of his time and toward the remarkable advances in science and technology that have emerged over the last 400 years. That progress has involved a symbiotic, evolving relationship among the creative minds of individuals, tools and practices for observation and experiment, social practices for systematic feedback on received knowledge, market and non-market systems for generating, disseminating and applying advances in knowledge, and finally, in recent decades, revolutionary information technologies that empower the human mind by expanding its limited capacities for raw information processing.

Turning from Francis Bacon in the 17th century to Karl Popper in the 20th, we find a convergence of thinking between them on the relationship between the human mind and external

---

[137] Gaukroger, S., *Francis Bacon and the Transformation of Early Modern Philosophy*. Cambridge University Press, 2001 (pp.10, 14–18); Kors, A., "The New Vision of Francis Bacon," Lecture 3 in *The Birth of the Modern Mind: The Intellectual History of the 17th and 18th Centuries* (recorded lectures from The Great Courses).

[138] Bacon, F., *Novum Organon* (1620), Note 13, Aphorisms No. 41–44.

tools. Popper saw ideas existing in three realms or worlds.[139] As a primary example, consider ideas expressed in books. The books themselves exist in World 1, the world of physical things, where "a book" may have multiple copies, paper and electronic. When we read a book, our various, conscious, subjective understandings of what we read exists in World 2, the world of mental states. That which multiple copies of a book all have in common—their text, their contents—exists in World 3. Note that an idea in World 2 might vary somewhat from one person to another, while an idea in World 3 is not variable in that way.

Once created, the contents of World 3 exist as objective knowledge, independently of the physical and mental versions in Worlds 1 and 2. So "World 3 is autonomous: in this world we can make theoretical discoveries in a similar way to that in which we can make geographical discoveries in World 1."[140] By that learning process, World 3 changes World 2, which may in turn lead us to change World 1. The three worlds thus interact.

Moving from World 2 to World 3 is central to human advancement. Science in particular depends on bringing knowledge from World 2 to World 3.[141] Doing so creates new opportunities to access knowledge, test it, and apply it to human needs. These activities foster an evolutionary selection process, with errors and new knowledge coming to light.

Consider technologies like the printing press and the computer, and simple practices like recording data at the time of observation instead of relying on unaided memory—they are powerful because they accelerate the movement from World 2 to World 3.

Yet medicine is mired in World 2. Misguided dependence on the human mind is built into the doctor's role. Changing that role would permit a radically new division of labor. And changing the division of labor would enable transforming the quality and economics of medical practice. But for this to happen, for moving from World 2 to World 3 to be fully effective, new information tools and new roles for their users are needed.

And everyone needs to understand why and how. By thinking in terms of World 2 and World 3, we can see more clearly how the doctor's current role distorts the division of labor, how a new division of labor could evolve, and how psychosocial, economic, and legal barriers are blocking that change.

---

[139] "We can call the physical world 'world 1,' the world of our conscious experiences 'world 2,' and the world of the logical contents of books, libraries, computer memories and suchlike, 'world 3.'" Popper, K., *Objective Knowledge: An Evolutionary Approach* (Oxford University Press, 1972), p. 74.

[140] *Objective Knowledge*, p. 74. Readers familiar with Plato may notice similarity between the Platonic theory of forms and Popper's World 3. The difference is that World 3 is a human creation, while Plato conceived the world of forms as a pre-existing reality.

[141] See *Medicine in Denial* (see Note 4), pp. 105–111, and Section 2.1.

6.2   THE DOMAIN OF COMMERCE: ECONOMY OF KNOWLEDGE

The philosopher Alfred North Whitehead observed:

> It is a profoundly erroneous truism … that we should cultivate the habit of thinking about what we are doing. The precise opposite is the case. Civilization advances by extending the number of important operations which we can perform without thinking about them.[142]

F. A. Hayek found Whitehead's principle to have "profound significance in the social field." Its significance lies in economy of knowledge: "We make constant use of formulas, symbols and rules whose meaning we do not understand and through the use of which *we avail ourselves of the assistance of knowledge which individually we do not possess.*" Hayek goes on to discuss "the unavoidable imperfection of man's knowledge and *the consequent need for a process by which knowledge is constantly communicated and acquired*" (emphasis added).[143] Hayek's concern was "the price system as a mechanism for communicating information." He observed:

> …the most significant fact about this [price] system is the *economy of knowledge* with which it operates, or *how little the individual participants need to know* in order to be able to take the right action. In abbreviated form, by a kind of symbol, *only the most essential information is passed on and passed on only to those concerned.*[144]

Just as market economies need a price system to efficiently communicate the limited information essential for individual transactions, so patients and practitioners need an efficient system for accessing and processing the limited, personalized information in a maximally usable form (i.e., trustworthy, timely, comprehensible information that is organized for decision making and therefore actionable). Like price information, personalized medical information creates stronger market incentives than general knowledge by facilitating accurate decision making on issues of personal importance.

But personalized information is a needle in the haystack of medical knowledge and data. Patients thus face enormous uncertainty unless and until they can access the limited information relevant to their individual problems. Resolving this uncertainty for patients is the traditional role of expert doctors.[145] But consumer dependence on costly experts for personal decisions interferes with the market, blocking the efficiencies that market forces otherwise tend to achieve. By compari-

---

[142]   *Medicine in Denial* (see Note 4), p. 107, quoting Whitehead, A., *An Introduction to Mathematics,* 1911 (American ed., Oxford U. Press 1948), p. 83.

[143]   F.A. Hayek, "The use of knowledge in society," *Am. Econ. Rev.*, XXXV, No. 4, September 1945, pp. 519–30 at p. 525 (emphasis added).

[144]   Ibid. at 522.

[145]   Arrow, K., Uncertainty and the welfare economics of medical care. *Amer. Econ. Rev.* 1963. LIII:941–73. See Note 121 for a reference to our further discussion of Arrow.

son, market forces would respond to economic reality—the value offered by external tools. The right tools are more efficient and effective than the minds of experts for accessing the limited information relevant to unique individual problem situations.

In many economic contexts other than health care, we take for granted that personal consumption decisions do not require costly expert advice. One need not hire an engineer to buy a car; one need not hire a professional guide to determine the route for driving through an unfamiliar area. Market and regulatory forces have developed tools and infrastructure (maps, road signs, and electronic navigation, for example) enabling consumers to function autonomously in such contexts. Rather than rely on third party agents to make group decisions, consumers act autonomously to make individualized decisions. But the complexity of medicine requires tools and systems for both consumers and professionals to cope with the complexity.

In medicine, tools for reliably processing complex information can *simplify* the ultimate choices presented to consumers. The tools filter out what is truly extraneous while presenting individually relevant options and the pros and cons of each in detail.[146] Without a system for accessing that information as needed, patients will continue to rely on the apparent expertise of practitioners. In turn, absent the necessary system, as Chris Weed has written:

> …practitioners might just as well continue to rely on their own creative intuition, experience, and random and informal contacts with other concerned people. Without the routine use of powerful knowledge coupling tools to generate specific linkages of the knowledge base to practical decision-making for unique individuals, scientific medicine affects practice primarily through new procedures and associated technologies, while the application of such procedures and technologies is left to a sort of cottage industry or folk art based on something approaching oral tradition."[147]

In short, both clinicians and consumers are unable to cope with complexity when left to their own devices. Both clinicians and consumers need to rely on external systems to manage information. Moreover, they need to use these systems jointly. These systems must therefore be simple to use for everyone involved. Indeed, simplicity at the consumer level is characteristic of much economic activity outside of health care. "The growing complexity of science, technology and organization does not imply either a growing knowledge or a growing need for knowledge in the general population," as Thomas Sowell has written. "On the contrary, the increasingly complex processes tend to lead to increasingly simple and easily understood products. … Organizational progress parallels that in science and technology, permitting ultimate simplicity through intermediate complexity."[148] From this perspective, the health care system's impenetrable complexity is anomalous.

---

[146] "Opening the black box of clinical judgment" (see Note 121), Part IV, pp. 7–8.
[147] Weed, C. C., "Overview," in *Knowledge Coupling* (see Note 18), p. xviii–xix.
[148] Sowell, T., *Knowledge and Decisions* (Basic Books, 1980, 1996), pp. 10–11.

Analysis by the Institute of Medicine (IOM) points in the same direction—simplicity must be built into the health care system for patients and practitioners. The IOM cites a theoretical basis for this conclusion in the study of "complex adaptive systems." Occurring in various social and natural contexts, complex adaptive systems are not built according to external, pre-conceived designs. Rather, complex systems "can emerge from *a few simple rules* that are locally applied" by individual participants in the system (emphasis added).[149]

What are the "simple rules" needed by the health care system? A basic reality of health care is its information-intensive nature. That reality suggests that simple rules for managing complex health information are pivotal. Consider an analogy from the domain of commerce: accounting rules for managing complex financial information. Accounting rules are complex. But that complexity exists only at the margin. The core concepts of double-entry bookkeeping are so simple that they are taken for granted. They apply universally, and yet allow for enormous diversity. They help to organize the economic relationships among individuals who may or may not have any awareness of them.

Something like this is needed in medicine. It can be found in the core concepts of problem-oriented care, and in the more detailed, but still simple to comprehend, standards and corresponding tools for informational process guidance, as described in Chapters 7–9.

---

[149]   IOM/NAM, *Crossing the Quality Chasm* (Washington, D.C.: National Academies Press, 2001), Appendix B, Plsek, P., "Redesigning Health Care with Insights from the Science of Complex Adaptive Systems," pp. 313, 316, available at http://books.nap.edu/openbook.php?record_id=10027.

PART II: THE SOLUTION

… a whole calling may have unduly lagged in the adoption of new and available devices. It may never set its own tests. There are precautions so imperative that even their universal disregard will not excuse their omission.

—Judge Learned Hand[150]

[150] *The T. J. Hooper,* 60 F.2d 737, 740 (2d Cir. 1932).

CHAPTER 7

A Problem-Oriented System of Health and Health Care

In a complex social system, its participants must be oriented toward a common general purpose, a purpose that different individuals may pursue for their own specific needs. In medicine, that common general purpose is individualized problem solving. To pursue that purpose, patients and their clinicians need a reliable infrastructure of tools and processes and information and standards of care, with corrective feedback loops and enforcement[151] so that individual and collective actions remain oriented toward the common general purpose.

The examples in Chapter 3 illustrate some of the ways things go wrong when the needed orientation and infrastructure are missing. Take the endometriosis case in Section 3.1.1. There the patient's problem was to get relief from her pelvic pain. But she endured 12 years of futility before getting the relief she sought. Her doctors first needed to determine the correct diagnosis. But they were oriented toward merely going through the usual motions of diagnosis for their usual fees while taking no more than their usual time. They were not oriented to solving their patient's problem, and so stopped short of doing so (they accepted preliminary diagnoses as final and did not follow through with more hypotheses and testing). They just took the path of least resistance. Missing was an integrated system for accountability, oriented toward the best path for solving the patient's individual problem. The system should make that best path for patients become the path of least resistance for clinicians.

---

[151] The concept of enforcement, which this book mentions repeatedly, and the related concept of audit, are central to LLW's system of problem-oriented care. Mere adoption of guidance tools without audit or enforcement may itself bring about some quality improvement, because using the tools for guidance is sometimes the path of least resistance and most reward in the short term. But that is not enough. Continuous improvement depends on continuous feedback and action on what the feedback reveals, which means a continuous assault on the status quo.

The terms audit and enforcement suggest scrutiny and action by outside parties, but organizations also need internal audit and enforcement, as is done with financial accounting standards. For example, two pioneers of the problem-oriented record, Doctors Harold Cross and John Bjorn, developed a system for internal audit of their practice's medical records, to enforce collection of a pre-defined history-physical-lab data base and completion of a problem list consistent with the data base. See their book, *The Problem-Oriented Private Practice of Medicine: A System for Comprehensive Health Care* (Modern Hospital Press/McGraw-Hill, 1970). A PDF of this book is available at www.world3medicine.org.

The best path for problem solving must be individualized to the problem as it exists in each patient over time.[152] That may well differ from the problem as it exists in the clinician's mind (e.g., the preliminary diagnoses in the endometriosis case) or as described by the generalizations of medical knowledge (e.g., the "classic" or "textbook" case, or the population averages of evidence-based medicine). Sound medical decision making demands constantly coping with these discrepancies between patients' actual problems and what clinicians expect and what medical "knowledge" says. The discrepancies must be taken into account, because they may well be crucial to making the best decision.

7.1   DEFINING INPUTS WITH GUIDANCE TOOLS

We began this book by asserting that a primary root cause of medicine's dysfunction is misguided dependence on the doctor's mind, operating in Popper's World 2. The mind's inputs for decision making are medical knowledge, patient data, and cognitive processing of that information. Those elements are undefined and uncontrolled when the mind operates autonomously, without external tools and standards of care. That leaves decision making incapable of organized, continuous, reproducible improvement. But this state of affairs is transformed when we move from World 2 to World 3. In World 3, inputs may be explicitly defined and brought under control.

To some, the notion of defining or controlling the mind's activities is anathema. The notion suggests conformity to external restrictions by exclusion of relevant information and suppressing clinical judgment. The simplistic "cookbook medicine" of managed care guidelines was criticized on this basis. Similar criticism—for failure to take into account evidence about individual variation—is leveled against "evidence-based medicine" guidelines derived from randomized clinical trials and large population studies. From these perspectives, the notion of defined or controlled inputs and processes sounds like a pseudo-scientific euphemism for compromising professional autonomy and the "art of medicine." Health IT threatens to become an insidious mechanism for imposing external controls over clinicians and clinical practice. And now AI threatens to displace the human role.

This skepticism reflects more than recent experience in medicine. It reflects also a broader critique of formal, rule-based approaches to expert decision making in many fields. In medicine this critique has been directed at practice guidelines, statistical decision analysis and computer-based tools. As summarized in a study by Marc Berg, this critique idealizes the "art of medicine" and doctor autonomy.

> Decision-analytic techniques … are but poor representations of the complexities that go into real-time decision making. One cannot separate the decision from its context … Such rigid, pre-determined schemes [as protocols] are said to threaten the physician's "art" by dehumanizing the practice of medicine and by reducing the phy-

---

[152] For further discussion of this concept of individualized pathways, see *Medicine in Denial*, pp. 51–52.

sician to a "mindless cook" … Moreover, such tools open the way for increased and uninformed controls by "outsiders." … All in all, these critics argue, the tools' impoverished, codified versions of physicians' know-how do not do justice to the intricate, highly skillful nature of medical work. The idea of creating formal tools that make medical decisions is utterly mistaken. Every attempt to take practical control of the decision process out of the physician's hands is doomed to fail—and is dangerous.[153]

Doctors' skepticism of rule-based controls over decision making is warranted, but their idealizing the "art of medicine" is not.[154] Doctors are right to condemn forms of control that involve exclusion of information and external power over decision making. But doctors are in denial about the extent to which they themselves impose such controls on patients. Doctors are right to reject impoverished, cookbook medicine, but they are in denial of how impoverished their own know-how is. So too are they in denial when they view themselves as "highly skillful," because their demonstrable skill levels would be higher if they used the right tools within a disciplined system of care. Physicians are right that "one cannot separate the decision from its context," and they are right to reject "uninformed controls by 'outsiders.'" But they are in denial of how much they themselves are uninformed outsiders to patients' lives, outsiders whose exercise of control inevitably separates medical decision making from its context. And they are in denial of the need to submit to different forms of control over their own inputs to care—both decision-making inputs and execution inputs.

Control need not mean exclusion of information or suppression of judgment. Instead, control should simply mean setting high minimum standards for *inclusion* of information and requiring a tool-driven approach for assembling that information. Such control leaves freedom to *exceed* the minimum with further information judged relevant by clinician or patient. But that further information must be defined, that is, made explicit as part of the basis for decision making.

Inputs to decision making (knowledge, data, and processing of that information) may be judgment-driven or tool-driven. If judgment-driven, we cannot know exactly *what* information clinicians take into account, nor can we know *how* they take it into account, nor can we reliably improve the cognitive processes involved. All we really know is that the basis for decisions is enormously variable. A patient has no assurance that different clinicians will converge on the one best practice for that patient's problem.

Remedying these failures requires: (1) high standards of care for managing health information (knowledge and data); (2) information tools designed to implement the standards in real-world practice; and (3) accountability for adherence to the standards via habitual, disciplined use of the tools and ongoing scrutiny of what the tools expose.

---

[153] Berg, M., *Rationalizing Medical Work*. (MIT Press, 1997), p. 7.

[154] "We debase the word "art" itself when we call what we've been doing art. And it's not science." These words come from LLW's discussion of "the art of medicine" at the end of a video of a 1971 grand rounds presentation, cited in Note 129.

But how are the tools and standards to be designed? What is their organizing principle? The answer to those questions lies in the concept of problem orientation.

7.2   THE MEANING OF PROBLEM-ORIENTATION

A couple of decades ago it became conventional wisdom to say that health care should be "patient-centered." A related concept was that health care and its financing should both be "consumer-driven." These concepts (we will use "patient-centered" to refer to both) emerged as a reaction against traditional, provider-driven medical practice and then against payer-driven managed care.

From medical, economic, and humanitarian perspectives, this movement toward patient-centered care is absolutely headed in the right direction.[155] The right direction becomes still clearer when we recognize that health care is part of a larger health domain. In that domain, health is not a just technical field for expert clinicians and not just a commercial endeavor by institutional providers. Health is a universal pursuit by ordinary people, each of them unique in their physiology, psyche, and circumstances. Their circumstances include social drivers of health outside the health care system's control. Their uniqueness gives each of them personal expertise, no less important than the professional expertise of their clinicians. As LLW often said, each person has a de facto Ph.D. in personal uniqueness, meaning long-term, lived experience with one's own medical individuality. And each has a personal stake and motivation shared by no one else.

At the same time, an increasingly large part of the health domain has become the health care system itself. That system includes advanced scientific knowledge and technology with a central role for expert professionals. Health care is thus co-produced by patients and clinicians, with their personal and professional spheres of expertise.[156] So patients, whether they intend to or not, inevitably act as one of their own providers (prosumers), and they do so in both their health care activities and their other health-related behaviors in the surrounding health domain.

Much of the health domain (social drivers of health and patients' autonomous behaviors) can only be influenced, not designed or controlled, by the health care system. The patient's pivotal role throughout the health domain mandates a patient-centered design for the health care system portion of that domain.

For example, when a patient first sees a clinician with a medical problem for diagnosis or treatment, a patient-centered approach should mean that initial data collection and analysis does not vary from one clinician to another. Yet, those front-end activities are typically not patient-cen-

---

[155] See *Medicine in Denial*, part II.B.2.d (pp. 49–52, entitled "Consumer-driven spending and consumer-driven care") and Appendix A (pp. 253–256, entitled "Scientific principles that tell us why people must manage their own care").

[156] Batalden, M. et al., Coproduction of health care service. *BMJ Qual. Saf.*, 2016;25:509–517. DOI: 10.1136/bmjqs-2015-004315. Dr. Charlie Burger informed me of this article. George Reigeluth, founder of Prosumer Health, informed me of the term "prosumer," which comes from Alvin Toffler.

tered. Instead, they vary enormously, depending on which clinicians the patient sees and in what sequence. This front-end variation is driven by clinician idiosyncrasies (including specialty orientation)—the opposite of a patient-centered approach, which should be driven by individual patient needs. In particular, patients need to have all medical specialties taken into account up front. Failure to do so is a recipe for disorganization, waste, and error.

Individual needs continue to be neglected at the back end, when treatment decisions are made. At that stage, rather than carefully individualizing treatments over time, clinicians can easily overlook unique individual needs. Instead, they fall back on limited personal knowledge, or customary local practice, or one-size-fits-all dictates of third-party payers, or the marketing messages of drug and device vendors, or "evidence-based" guidelines derived from large population studies.

A truly patient-centered system would take precisely the opposite approach. The front-end investigation would not be variable but standardized (i.e., without case-by-case judgmental shortcuts), which is the only way to assure that the investigation is complete for every patient. The standardized data should be highly detailed, which is the only way to capture the patient's uniqueness.

Then, at the back end, decisions could become individualized based on the detailed data collected. This would lead to variable decisions for different patients labeled with the "same disease." Variation would be justified to the extent that the same disease label hides individual differences in medical need. The differences would be reflected in detailed data showing how the "same disease" is *not* the same for different patients. Stated differently, the disease label is an abstraction from the actual disease as it variably exists in different individuals. Yet, the "same disease" abstraction may still be meaningful. It may capture scientific knowledge about crucial commonalities among the different individuals (for example, all diabetics have in common some dysfunction in hormonal regulation of blood glucose levels). That scientific knowledge as manifested in each individual must be taken into account—which can only happen when detailed data on that individual are coupled with medical knowledge about what the data mean.

The foregoing describes the process needed for handling a single problem. Most patients have multiple problems. So the same process is needed for handling each one of those, taking into account how the various problems and medical interventions interrelate. All of this must be recorded in EHRs that capture what is going on with each problem in light of the others over time.

So patient-centered care must be driven by the needs of unique individuals in all their complexity. From this perspective, "patient-centered care" is but a general ideal. The ideal must be translated into a specific, operational approach for coping with complexity. That approach begins with analyzing complex needs in terms of specific problems and orienting patient care activities toward solving those problems.[157] This orientation must be simultaneously reductionist and holistic. The total needs of the "whole patient" should be broken down into specific, identifiable problems. Activities to solve each problem should be carried out in light of the other problems. Both medical knowledge and the patient's personal knowledge should inform the collective activities of multiple clinicians and the patient.

In contrast to a problem-oriented system of care, the current non-system lacks any clear orientation at all. The non-system is pulled in different directions by the beliefs and habits and financial interests and power relationships of the people and organizations involved, with little accountability for any of them. Such a non-system inevitably runs wild, medically and economically.

To create accountability, a problem-oriented health care system must satisfy three general principles.

- All of the individual's health-related problems must be taken into account, not limited by the personal knowledge or interests of clinicians or third parties.

- Problems must be handled in a structured, step-by-step way, following the basics of orderly problem solving in any field, driven by the realities of each the patient's problems as they actually exist in the unique patient.

- Problem solving must have scientific integrity, for which guidance and accountability mechanisms must be built into the problem-solving structure, to assure reliability and effectiveness.

Now recall the point made in Section 1.2: everything depends on equipping people with the right tools. From that perspective, we can translate the above three general principles into three core requirements for problem-oriented health record tools. Those requirements and a detailed discussion of the record are presented in Section 9.1.

But it turns out that this process guidance via the record of care is not enough. Informational inputs to the record are traditionally determined by the doctor's mind. The doctor decides what hypotheses to investigate, what data to collect, and what the collected data mean in light of the

---

[157] See Popper K. *All Life Is Problem Solving* (Routledge, 1999), p. 100, where Popper describes problem solving as a trial and error process, which occurs in both biological evolution and social evolution (in the latter, the process may or may not be conscious and designed). As to social evolution, LLW observed that medicine has failed to consciously design social and technical processes for improvement by systematic trial and error:

> We ourselves are the agents of our own extrasomatic evolution. Physical man has changed little over thousands of years. It has been his tools and beliefs that have enabled our civilization to develop. The present premises and tools of medical education and medical care cannot support or even allow the proper evolutionary steps to take place.

Knowledge Coupling (Note 18), p. 5. In that book and in a 1981 article (Note 17), LLW examined how to design a new division of labor, enabled by a new informational infrastructure, to allow the proper evolutionary steps to take place. See also the article quoted at Note 123.

personal knowledge of the doctor or a consulting specialist. Yet, no doctor's mind can cope with the necessary coupling of vast knowledge with detailed data. Burdening the mind with this task necessarily falls short of truly patient-centered care. Relieving clinicians from that burden means that an EHR must be used in conjunction with a CDS tool optimized for the task and designed for joint use by patients and clinicians.[158]

Together, the EHR and CDS tools form part of the infrastructure for individualized health care delivery and knowledge development. See the image below, providing a diagram of this infrastructure, based on the figure on p. 13 of *Medicine in Denial*.

Individualized Healthcare Delivery and Knowledge Development Systems

___

CHAPTER 8

Informational Guidance: Clinical Decision Support Tools and Standards of Care for Coupling Patient Data with Medical Knowledge

8.1 THE TWO STAGES OF DECISION MAKING

Recall from Section 2.1 that decision making can be conceived in two stages: (i) assembling information to take into account; and (ii) making the decision, based on that information, preferences, and values. The first stage includes preliminary decisions about what information needs to be assembled. These preliminary decisions should be largely tool-driven, and should not be confused with ultimate decisions in the second stage, which should be reserved for human judgment.

Ultimate decisions in the second stage depend on the informational foundation laid in the first stage. It follows that the roles assigned to the human mind and external tools in the first stage are foundational for all health activities driven by decision making, whether preliminary decisions in the first stage or ultimate decisions in the second.

The roles assigned to the human mind and external tools in the first stage are the primary subject of this chapter. Before addressing that subject, however, we need to briefly discuss the second stage.

In the second stage, the question is, who decides—the patient or the expert clinician? Or should both of them submit to decisions arrived at by a third party or external AI tools? The answer is that the patient's judgment should govern the second stage. This becomes apparent by viewing decisions in one of two categories.

- Decisions where the assembled information makes the best option so obvious that no real judgment is required—the same decision would be made no matter who decides. This category of decision might be labeled "information-driven."

- Decisions where more than one option is medically plausible, difficult trade-offs are unavoidable, and uncertainty thus remains even after all relevant information is taken into account. Because this category of decision is sensitive to patient preferences among competing values, this category might be labeled "values-driven" or "preference-sensitive."

An example of an information-driven decision is the case described in Section 3.1.3.1—the girl with Addison's disease. Once the right information was assembled, the right diagnosis became "obvious," and the right treatment for that disease was clear. So the advantage of regarding the patient as the decision maker in these information-driven cases is that obtaining patient's assent makes the patient more informed and more committed to the decision—which especially matters if the patient is involved in executing that decision. The first stage of decision making should be designed to maximize the number of such cases, by assembling the right information.

The second category is where patient judgment must govern. In these situations of genuine uncertainty, the patient has the greatest stake in solving the problem, the patient must live with an unsolved problem, the patient has first-hand experience of his or her unique set of personal characteristics and circumstances, the patient must bear the risks and benefits and burdens and trade-offs involved in deciding among uncertain options, and the patient must become informed and motivated to execute decisions where the patient is the primary actor (as with lifestyle and behavioral changes, medication adherence, physical therapy).

The patient may forego exercising authority over the second stage by deciding to defer to the judgments of a trusted clinician or group of clinician colleagues. But that deference should result from the patient's choice, not from paternalistic authority assumed by clinicians. Patient deference to clinicians is not optimal as a general matter, because it creates risk of an unhealthy dependency relationship and does not foster patient understanding, feedback, and commitment. Nevertheless, deference to the clinician may be reasonable in situations such as the following:

- where the pros and cons of a decision are mainly technical issues that don't implicate the patient's preferences and values (for example, deciding between two drugs that have different mechanisms of action but are otherwise comparable in terms of individualized risks, benefits, and costs);

- where the clinician has a close relationship to the patient, plus long experience with many other patients in similar situations, such that the patient trusts the clinician to make the best decision from the patient's perspective; and

- where the patient does not feel up to sorting out the best decision and has no family or advocate other than the clinician.

In such situations, the patient who defers to clinician judgment should understand and accept the risk that the clinician could misjudge the patient's interests and desires. That risk is significant, because no clinician shares the patient's self-knowledge, life experience, and living circumstances. Moreover, clinicians are burdened with many patients and can't necessarily develop an in-depth understanding of each one.

Now let's return to the first stage of decision making. Analyzing that will provide background for considering the foundational issue raised at the beginning of this section: the roles assigned to the human mind and external tools. Then, in Section 8.2 we explain alternative approaches to the first stage of decision making.

The first stage begins with the initial workup of an identified problem. (The problem is identified when a patient experiences a symptom and sees a clinician, or when a clinician uncovers some abnormality, for example, upon doing a physical exam.) The initial workup of the problem can be conceived in four phases—choice, collection, and analysis of patient data, followed by output of the results.

- **Choice:** Initial data must be selected for identifying and assessing diagnostic or therapeutic options. This begins with linking one data point—the patient's problem—with medical knowledge about (i) potentially relevant options and (ii) data needed to determine which options could be worth considering for this patient.

- **Collection:** All the data chosen in the preceding phase must be collected and properly recorded without error or omission.[159]

- **Analysis:** Once collected, the chosen data must be linked with comprehensive medical knowledge to determine what combinations of data points are significant and what those combinations mean.

- **Output:** A set of options (diagnostic or therapeutic) worth considering for the patient, plus, for each option, evidence (patient data) for and against the option. This core output should be accompanied by explanatory medical knowledge about the options and evidence plus additional background to consider.

These four phases apply to both diagnosis and treatment decisions. That both types of decision have the same logical structure is not surprising. Organized problem solving of any kind involves assembling information to identify options for how to proceed, plus pros and cons of each option. The resulting output from the first stage of decision making provides an organized basis for making the decision in the second stage.

---

[159] "Properly recorded" means that the data must be entered in the CDS tool providing guidance on what data to collect. That tool should be integrated with the patient's EHR and/or PHR, so that the data are also recorded in that record in the right category, which provides context. For each finding, not just the content but certain ancillary data should be recorded (e.g., a blood pressure entry should indicate whether the patient was at rest or exercising when the finding was made). This kind of detail is needed for data to be interpretable and "trendable" relative to other entries. See the blog post cited in Note 36.

These four phases are a high-level overview. They must be further specified as a seven-step sequence, presented in Section 8.3. That sequence provides a basis for software design.

Now let's return to the foundational question raised at the beginning of this section: what roles should be assigned to the human mind and external CDS tools in the first stage of decision making? The answer is that external tools should be the primary vehicle for carrying out that first stage—assembling information (knowledge and data). The human mind's role should be to *supplement* external tools in the first stage (and to *govern* the second stage, as already discussed).

Stated differently, the human mind should have freedom to add to, but not subtract from, tool-driven output in the first stage. This means that external tools should be used habitually, not selectively, and that the data collection indicated by the tools should be done completely, not partially (with very limited exceptions noted below). So use of external tools should be enforced to achieve the high standards defined by the tools while preserving freedom for human judgment to add information not elicited by the tools. These judgments, if validated, can be incorporated in the tools as part of an evolutionary feedback process of continuous improvement.

If the first stage is completed and carefully recorded, then, in the second stage, human judgment is fully informed and its basis transparent. The objective quality of the resulting decision, and the subjective trust of the parties involved, both depend on using the CDS and medical record tools faithfully.

To support these conclusions, the next section compares the tool-driven approach just described with the traditional judgment-driven approach to carrying out the above four phases of the initial workup. That comparison is followed by a more complete description of the tool-driven approach.

8.2 ALTERNATIVE APPROACHES TO THE FIRST STAGE

8.2.1 THE JUDGMENTAL APPROACH

The culture of medicine, and medical education in particular, have never clearly distinguished between the two stages of decision making. So medical schools have never taught doctors a systematic, scientific approach to assembling the informational basis of decisions. Instead, this assembly process has been left to each doctor's judgment (which involves both analytic thinking and intuition). And now patients are using Internet searches to add their own information assembly and judgments to the mix.

The first stage of decision making may begin before patient-clinician encounters. The patient experiencing an unexplained medical problem starts by judging what symptoms should be searched for on the Internet. Then the patient starts the encounter with the clinician by judging what to say about the problem. Then the clinician starts interrupting with questions almost as soon as the patient begins talking.[160] After hearing answers, the clinician judges what further questions to ask, and so on in a cycle. Inevitably, this cycle activates the frailties of human judgment and runs up against the limits of personal knowledge. Even the most brilliant and disciplined clinicians cannot fully escape these constraints.

In repeatedly relying on fallible knowledge and judgment, the clinician largely determines what gets discussed, what portion of the discussion gets entered in the medical record, and what gets overlooked. Patients may "wait until the last moment in the clinical encounter—often while the physician is grasping the doorknob to exit the examination room—to utter something that, not uncommonly, provides crucial information."[161] This doorknob phenomenon is but one example of how a judgmental approach tends to be ad hoc, hit-or-miss, incomplete, and not patient-centered.

Clinician judgment is not only fallible but idiosyncratic. Two clinicians starting with the same patient often go in very different directions, depending on what questions they first ask, how they analyze the patient's answers, and how they follow up. Practice guidelines are not sufficient to overcome these idiosyncrasies (see Section 3.3.1).

Moreover, the entire process is not cumulative and is thus inherently inefficient. Clinicians are constantly reinventing the wheel. Rather than accessing knowledge and analysis compiled *before* the patient encounter for later use by all, clinicians are instead constantly using their own minds to recall and apply medical knowledge anew *during* the patient encounter. This inefficiency is not only uneconomic but unscientific. The philosopher Alfred North Whitehead observed that an activity constantly requiring "a fresh display of ingenuity … lacks the great requisite of scientific thought, namely, method." A method or system maximizes use of existing knowledge and analysis. This is consistent with Whitehead's further point that "Civilization advances by extending the number of important operations which we can perform without thinking about them" (see Note 141).

The need to "perform without thinking" brings us back to the frailties of human judgment. Performance without thinking protects against those frailties, as well as making productive use of scarce resources (time, attention, money).

The first stage of decision making presents another difficulty. Medical knowledge itself is distorted by the human mind, as discussed in Section 3.5. Knowledge is usually expressed as population-based generalizations that fail to capture countless individual variations. But properly applying the generalizations requires also taking into account the variations, which in turn requires collecting detailed, patient-specific data.[162]

---

[160] "The average patient visiting a doctor in the United States gets 22 seconds for his initial statement, then the doctor takes the lead." Langevitz, W., Denz, M., Keller, A., et al., "Spontaneous talking time at start of consultation in outpatient clinic: cohort study." *BMJ*, 2002; 325:682-69. DOI: 10.1136/bmj.325.7366.682.

[161] Faden, J. and Gorton, G., The doorknob phenomenon in clinical practice. *Am. Fam. Physi.*, July 1, 2018;98(1):52-53.

[162] See generally *Medicine in Denial*, part VII.

This does not mean that human judgment has no place. On the contrary, human judgment can usefully supplement a tool-driven process for the first stage of decision making. And, as the preceding section makes clear, human judgment is essential when the second stage of decision making involves genuine uncertainty (where the best solution is not determinable from medical science), as distinguished from mere personal uncertainty in Popper's World 2 (where a clinician happens to lack World 2 personal knowledge of an optimal solution determinable from World 3 medical science).

8.2.2 THE COMBINATORIAL APPROACH

"There remains but one cure, one healthy course: the whole operation of the mind must be completely re-started, so that from the very beginning it is not left to itself but is always subject to rule; and the thing accomplished as if by machinery."

—Francis Bacon[163]

Information tools external to the human mind enable various alternatives and supplements to the judgmental approach, depending on how the tools are designed and used. This section describes an alternative we refer to as a combinatorial approach. We use the term "combinatorial" simply as a shorthand for the general concept of systematically working through a finite set of elements to find significant combinations of those elements. Thus, we do not use the term in a strict mathematical sense. In medical practice, the elements to be combined are data points on a patient and knowledge about different combinations of data points with medical significance.

A judgmental approach relies on human judgment (both analytic and intuitive) to recall or recognize significant patterns in a mass of information. This task is described by metaphors like "connecting the dots," or "finding a needle in a haystack," or "separating the wheat from the chaff." Human judgment attempts this task using recall, scientific inference, and intuition, all of which are vulnerable to cognitive shortcuts (heuristics and biases).

By comparison, a combinatorial approach bypasses human judgment during the initial workup, relying instead on external tools for clinical decision support. These CDS tools are used for simple sorting and matching, exhaustively performed without shortcuts. Such simple tasks do not require advanced intelligence, human or artificial, but they do require the high speed and capacity offered by digital tools.

The patient begins the combinatorial process by using a CDS tool specific to the problem presented by the patient (for example, a tool for diagnosis of chest pain, or a tool for management of diabetes). The patient answers a meticulously detailed set of questions about the problem's characteristics and history and related circumstances such as risk factors. Then the CDS tool should guide

---

[163] Bacon, F., *Novum Organon* (1620), Note 13, Preface to Second Part.

the clinician on physical exam observations to make, basic lab tests to order, and further data that may be readily available from other sources (e.g., social determinants of health or genomic data). Collecting richly detailed data in this way begins the initial workup—the threshold of the first stage of decision making.[164] Using a CDS tool enables searching countless combinations of data points for those that are medically significant. The significant combinations will suggest more options than any one clinician would think of.

The detailed data collection exploits the power of simple, non-specific findings often to become highly specific light of medical knowledge when they appear in combination. The combinations occur at two levels: (i) combining all collected findings with medical knowledge, resulting in (ii) combinations of findings (finding sets) known to be medically significant. Recall, for example, the Addison's disease case in Section 3.1.3.1. The correct diagnosis could have been considered and confirmed at or near the outset of care had someone "connected the dots"—that is, recognized that some of the many early findings showed a pattern characteristic of Addison's. No one finding was specific to the disease, but their appearing in combination was highly specific. Symptom combinations are not always that specific (for example, early COVID-19 symptom combinations), but they are still useful for comparing the full range of diagnostic possibilities.

For such combinations to be readily seen, the CDS tool should present each significant combination of findings (positive, negative, and uncertain), identify the diagnostic possibility it suggests, and provide explanatory material (see Sections 8.3.3.6 and 8.3.3.7 for an example). This is the output that the combinatorial approach generates.

In contrast, the judgmental approach runs off the rails as soon as the clinician and patient start talking. They both implicitly judge what data need to be considered first; then the clinician judges what the initial data mean and what further data collection is needed, going through the Q&A cycle described above. Neither of them should attempt those initial judgments (but both should be free to add their personal judgments later in the process).

Combinatorial analysis thus transforms the first stage of decision making—assembling information—from a judgment-driven, idiosyncratic, error-prone, opaque, expensive process to a tool-driven, consistent, reliable, transparent, inexpensive process. Another difference is that the judgmental approach is quicker up front but prone to avoidable, time-consuming trial-and-error, while the combinatorial approach takes longer up front but protects against time-consuming blind alleys and crises later. These sharp contrasts occur because the judgmental and combinatorial approaches differ radically in how and when they incorporate human judgment.

- The judgmental approach relies on the clinician's judgment to apply personal knowledge from Popper's World 2 on the fly *during* patient-clinician encounters. The combinatorial approach instead first operates *before* clinician-patient encounters by relying

---

[164] See *Medicine in Denial*, part IV.A (pp. 53–56), distinguishing between "threshold processes" and "follow-up processes," and explaining why the former is especially important.

on World 3 objective knowledge developed by careful judgments from leading experts and incorporated in external tools. Those tools are then used as needed by patients and clinicians before, during and after their encounters.

- The judgmental approach is a black box, being variable and obscure in what patient data it takes into account and how it does so. The combinatorial approach is consistent and transparent in using knowledge to identify (i) detailed patient data to be collected and (ii) medically significant combinations of the collected data points.

- The level of detail, in both data collected and knowledge taken into account up front, is far greater with the combinatorial than with the judgmental approach. This greater detail brings out the uniqueness of every patient and every problem situation. The generalizations of medical knowledge and the frailties of human judgment hide that uniqueness and thus compromise individualized decision making.

- The combinatorial approach processes this detailed information up front and all at once. In contrast, the judgmental approach is gradual and piecemeal. The clinician first collects skimpy data and jumps to initial conclusions, and then repeats data collection and analysis in a cycle.[165] Three basic pitfalls result: information needed at the outset might not be gathered until later, if at all; premature judgments are made; and follow-up actions go all over the map without zeroing in on the best solution.

- In the combinatorial approach, the data collected up front include only readily available, easily collected findings. This limiting principle excludes findings that are invasive, painful, risky, costly, time-consuming to make, or unlikely to be productive at the initial workup stage. In contrast, the judgmental approach permits premature resort to such findings, before they are clearly necessary. An example is CT scans ordered before the patient is even seen—a routine occurrence that is rarely defensible.

- The judgmental approach fails to develop organized, continuous, rapid improvements and communicate those to patients and clinicians as needed during their activities. In contrast, the combinatorial approach uses external tools to rapidly harvest and distrib-

---

[165] Proceeding gradually and piecemeal has led doctors to apply "a method of statistical inference in which Bayes' theorem is used to update the probability for a hypothesis as more evidence or information becomes available" (Wikipedia, internal hyperlinks omitted). As applied in medicine, Bayesian inference starts with population-based probabilities on each option (usually failing to take into account how diseases evolve over time) and then modifies these with a few fragmentary details ("pretest likelihood") about the unique patient. As each new detail is introduced piecemeal, the probabilities for each option change such that prior probabilities quickly become irrelevant. The combinatorial approach avoids this failing by eliciting hundreds of data points and matching them with medical knowledge up front. For further discussion of the difficulties with Bayesian inference, see *Knowledge Coupling*, Note 18, pp. 38–40.

ute validated improvements from the medical literature and other sources. One source is supplemental data entries in free text and other feedback by real-world users of the CDS tools. Another source is researchers who study results of that use and other data as recorded in EHRs. In these ways the combinatorial approach fosters feedback and evolutionary improvement in real-world practice.

The foregoing comparison of the judgmental and combinatorial approaches focuses on inputs and internal processes. Another basis for comparison is outputs.

- The output of the combinatorial approach is not a decision or recommendation but rather the informational basis for a decision: a set of options with patient-specific evidence for and against each option. The options and evidence are generated by a CDS tool, not by the clinician's judgment. The exercise of judgment is left to the patient and clinician. In contrast, the output of a judgmental approach naturally takes the form of the clinician's recommendation plus a selective, incomplete presentation of options and evidence chosen by the clinician, who may be biased and whose authority may unduly influence the patient.

- The tool-driven combinatorial approach automatically generates complete documentation, which is available over time to the patient, any clinician who becomes involved, and, if authorized, researchers and regulators. This documentation includes all positive, negative, and uncertain findings, not limited to those the clinician judges significant. In contrast, typical clinician notes are very incomplete, and also ambiguous as to whether an unmentioned finding was negative or simply never made. Moreover, poor documentation inhibits effective research and regulation.

The combinatorial approach can be compared with algorithmic and probabilistic approaches. Both risk jumping to conclusions (premature closure) by excluding potentially relevant possibilities from full consideration.[166] Moreover, the combinatorial approach is sensitive to context in a way that makes it less error-prone and more fault-tolerant than algorithmic and probabilistic approaches.

Algorithmic approaches are usually rule-based, involving branching logic such as "if X is present, then do Y, not Z." But such algorithms are too simple and inflexible for complex, multifactorial medical decisions with many options. An algorithm might dictate one option without allowing for assessment in the context of other options and detailed data about the problem situation. Any algorithm determined in advance of patient-specific use should be treated as one of the options to be considered and compared with alternative options.

---

[166] See also *Knowledge Coupling: New Premises and New Tools for Medical Care and Education* (Note 18), pp. 38–40, 53–54.

Probabilistic approaches rely on population averages and probabilities. That reliance tends to divert attention from the diagnostic options most applicable to the individual patient. Use of averages and probabilities is "in direct proportion to our ignorance of the uniqueness of the situation."[167] So we should first reduce our ignorance by collecting detailed data about the situation.

Recall again the girl with Addison's disease from Section 3.1.3.1. As stated in the article (p. 48) about that case, "the clinician usually begins diagnostic investigation by considering (and excluding) the most common diagnoses." (Thus the saying among doctors—"when you hear hoofbeats, think horses, not zebras.") Prioritizing the "most common" alternatives meant that the uncommon diagnosis of Addison's disease "did not make the list [of diagnoses to consider] until it was nearly too late to save the child's life." Whether a disease is common or rare depends on the context. (So in the context of central Africa, the saying among doctors might become, "when you hear hoofbeats, think zebras, not horses.")

In the general population, Addison's disease is indeed rare. But the disease is common (perhaps almost universal) in patients with a combination of findings like the girl's fatigue, hypotension, weight loss, abnormal pigmentation, dehydration, nausea, and abdominal pain. People with a more-or-less similar pattern of findings (a tiny subpopulation) are the relevant context for determining what diagnostic possibilities are common or rare, probable or improbable. That context can only be determined by first collecting detailed, readily available data useful for assessing all the diagnostic possibilities.

Whether or not a disease is common or rare in the general population is irrelevant, indeed misleading, when first identifying what diagnostic possibilities are worth considering for a unique patient. Once those possibilities are identified, the common or rare status of the disease might become relevant to prioritizing further inquiry if several possibilities seem plausible (i.e., none of them stand out as the best fit with the initial detailed data). "But using population-based knowledge to select the highest priority diagnostic possibility is only an intermediate step. The next step is collecting new patient-specific data to confirm or rule out that possibility. Once collected, the new data supersede population-based evidence as a basis for decision."[168]

These same principles apply to deciding among treatment options. Recall the discussion of opioid treatment in Section 3.2.1. There we explained how alternative treatment options are like alternative diagnostic hypotheses. One option may be the "standard of care" or "treatment of choice" under a guideline. But that guideline is only a hypothesis that the favored treatment is in fact the best option for any particular patient.

---

[167] Ibid., pp. 12, 38, 54.
[168] *Medicine in Denial* (Note 4), p. 186. This point is from Chapter VII, entitled "The Gap Between Medical Knowledge and Individual Patients" (pp. 177–194). Among other things, that chapter further demonstrates why a tool-driven, combinatorial approach is essential for selecting and analyzing patient data.

Verifying that hypothesis requires comparing the other options to determine whether the favored treatment might in fact be inferior to some alternative treatment once patient-level details are taken into account. The patient and clinician need a tool informing them of details bearing on how each option would affect that patient relative to other options. In this way, decision making becomes oriented toward the actual problem as it exists in the unique patient (as distinguished from the abstract problem as variously conceived in the minds of the patient's clinicians and guideline authors).

Evidence-based treatment guidelines are generally based on population studies from randomized clinical trials. These trials typically involve carefully selected patients with little real-world variability (such as co-morbid conditions). Excluding variability to achieve randomization excludes differences that are often crucial to individualized decision making for real-world patients.

In short, the objective quality of decisions, and the subjective trust of the parties involved, both depend on using CDS tools to enforce scientific rigor in the first stage of decision making. This is what the combinatorial approach achieves. Consider what Dr. Ken Bartholomew has written of his real-world primary care experiences. With his patients, he used tools designed to implement the combinatorial approach (LLW's "coupler" tools for coupling medical knowledge with patient data). He found enormous value in tool-driven gathering of richly detailed data up front.

> My experience using couplers … has given me the gratification that patients' wants and needs as outlined above are continuously met by the couplers. Not only do the patients see the thoroughness involved in the use of couplers, but they sense that we care enough to give them the kind of thoroughness that they feel entitled to. With the coupler's systematic review of details in the patient's life that could be relevant to the current problem, the patient feels that his or her individual situation has been thoroughly examined and all possible conclusions have been taken into account.

Beyond a sense of thoroughness and caring, Dr. Bartholomew explains, patients also get an education in how to understand and participate in their own care.

> In management couplers, they further see the many different combinations of therapy and understand that the care of a complex, long term problem requires a detailed understanding of the patient's unique situation, followed by a careful monitoring of the options finally chosen. Even when a diagnosis is still in question, they have, in my experience, been completely satisfied with the outcome of the encounter. In addition, by receiving a printout of the findings and possible causes, they feel empowered to review the situation at home and to watch for signs and symptoms that may aid the diagnostic process in the days or weeks to come. The use of couplers teaches them that there is a time course to disease and not all signs and symptoms necessarily occur "by the book" or simultaneously. By thus empowering our patients with infor-

mation, as opposed to leaving them in a void, we reinforce their collaborative role as part of a team working toward an understood goal.[169]

Dr. Bartholomew goes on to describe his clinician perspective on use of knowledge coupling tools:

> Because of the timeliness of data that is built into the couplers, and since they are so pertinent by being problem oriented, I do not need to spend 20 or 30 minutes going through indexes of textbooks to find what may or may not be appropriate information.
>
> Using Problem Knowledge Couplers has, for me, become an enjoyable experience, now that I am comfortable with them. The couplers are full of information; I have never failed to learn something new each time I have used one. If this is true for an experienced clinician, think of the value that they would hold for medical and nursing students! Using couplers begins to become a reinforcing loop-because they are fun to use, you use them more. The more you realize that valuable information, beyond your own personal store of knowledge, is being brought to bear on each of the patient's problems, the more secure you feel, the more patient gratification you generate, and the more gratification you have from your practice.[170]

8.3 CORE OPERATIONS OF TOOLS IMPLEMENTING THE COMBINATORIAL APPROACH

8.3.1 OVERVIEW

To restate where we are, the first stage of decision making involves assembling the information that the decision maker needs to take into account. Recall from Section 8.1 that this first stage has four phases: choice, collection, and analysis of data, then output of the results. The first stage can also be broken down into three categories of options ("options" means diagnostic possibilities or therapeutic alternatives, depending on the problem), in preparation for the second stage where the best option is chosen. The following diagram depicts the three categories of options.

---

[169] Bartholomew, K., "The Perspective of a Practitioner" (Note 99), pp. 238–39.

[170] Ibid., p. 240. For further discussion of the operational specifics of using Couplers, see "Weeds' Responses to questions raised in Episode 20," http://imreasoning.com/wp-content/uploads/2016/12/Podcast-questions3-response-1.pdf, which is a written submission expanding on a September 2016 podcast interview with LLW. The interview is here: https://imreasoning.com/episodes/episode-20-medicine-denial-part-2-interview-larry-weed/, and an introduction by the podcast hosts is here: https://imreasoning.com/episodes/episode-19-medicine-denial-introduction-dr-larry-weed/. We are grateful to the hosts, Dr. Art Nahill and Dr. Nic Szecket, for inviting the interview and posting the written submission.

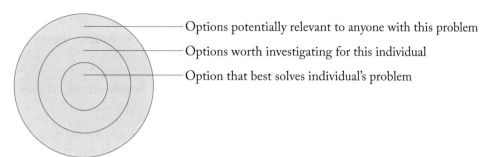

Options potentially relevant to anyone with this problem
Options worth investigating for this individual
Option that best solves individual's problem

The first stage of decision making can be further broken down into the following seven-step sequence.

1. **Link the initial data point (the patient's problem) with applicable knowledge.** That knowledge identifies all options potentially applicable to solving the problem as initially presented.

2. **Define each option as a set of simple findings from readily available sources.** Traditionally, these sources are symptoms/history questions to be answered by the patient, physical exam observations to be made by the clinician, and basic lab tests to be ordered by the clinician (some of this data may also be obtainable from prior medical records). Other sources are possible, however.[171]

3. **Combine all the finding sets for all the options from step 2 and generate a list of findings to be made.** Many finding sets will partially overlap in the items they include, which reduces the total number of findings to be made.

4. **Make the findings and record them as positive, negative, or uncertain.** Both the patient and clinician carry out this step. These responses may be supplemented with free text annotations.

5. **Match all recorded findings from step 4 with all finding sets from step 2.** This enables sorting the options into two groups: (i) options worth considering for this individual because they have at least one positive (or uncertain) finding; and (ii) options that can be safely ignored because all findings for the option are negative. Stated differently, this step takes the options of potential relevance to the problem in general and narrows them down to options of actual relevance to the problem as it exists for this patient in particular. Steps 6 and 7 involve only the options of actual relevance.

---

[171] These other data sources might include wearable medical devices, or public health agency data on social determinants of health, or genomic/molecular data that may have been previously collected and stored in some accessible repository.

6. **Generate initial output—a list of options, categorized in useful groups and, within those groups, ranked by the number of positive findings.** Alternatively, uncertain as well as positive findings could be included for ranking purposes.

7. **Generate detailed further output on each option.** This further output consists of knowledge useful for interpreting the findings and assessing the option.[172]

No clinician can perform all these steps anew in each case. Even attempting to do so would require pointless duplication of effort (recall Whitehead's point about avoiding fresh displays of ingenuity, discussed in Section 8.2.1). See steps 1, 2, and 7. These steps involve research that can be done in advance of any patient-clinician encounter. No clinician should duplicate that research (nor try to recall it) during patient care. Instead, a CDS tool should capture that research in distilled, actionable form. Then the tool can execute steps 1–3 and 5–7, with each step being virtually instantaneous. Only step 4 needs to be executed by the patient and clinician. This division of labor is crucial, from both medical and economic perspectives.

Immense human labor and careful, deliberate judgment are incorporated in the CDS tool for purposes of steps 1, 2, and 7. That labor occurs at two levels: (a) the original scientific and clinical research captured in the peer-reviewed medical literature and other trusted sources; and (b) the research conducted by medical content specialists who analyze and distill the literature for capturing its actionable points in CDS tools. At both levels, the researchers' output exists in Popper's World 3, where it becomes more reliable and usable by all, unlike knowledge residing in World 2, the minds of clinicians. In particular:

- the research involves collective human judgments reached under ideal conditions—the researchers have time to deliberate carefully within a system of feedback. These conditions protect against the cognitive vulnerabilities that plague human judgment under real-world time pressures; and

- knowledge residing in World 3 can be readily used by all who need it while being continuously improved and disseminated. Continuous dissemination of improvements is especially crucial in fast-moving areas like pandemic response and cancer therapy. By comparison, the status quo is that medical science advances take *17 years* to be adopted.[173] Delay of that magnitude results in large part from medicine's being mired in World 2.

---

[172] The above sequence of steps can be viewed as generic. This means it is not specific to medicine, nor to any particular software tool. Instead, the sequence is a process for matching general knowledge with specific data about a problem situation, for purposes of identifying options and evidence relevant to that situation.

[173] Parston, G., et al., The science and art of delivery: accelerating the diffusion of health care innovation. *Health Aff.*, Vol. 34, No. 12 (2015):2160-2166. DOI: 10.1377/hlthaff.2015.0406. ("It is widely acknowledged in health care that there is a lag of seventeen years, on average, between an invention or innovation and its widespread use across a health system.")

8.3.2   PROCESS FOR BUILDING CDS TOOLS TO IMPLEMENT A COMBINATORIAL APPROACH

In the next section, we describe the sequence of seven steps in much greater detail. Before doing so, however, we here explain how the CDS tools used to execute these steps were built. (This explanation is based on the operations at PKC Corporation, where LLW and his colleagues developed "knowledge coupling" tools, a form of CDS designed to implement the combinatorial approach, from the early 1980s to 2006.[174]) The tools have three components, containing both a software engine and extensive medical knowledge content. Specifically:

a. A specialized repository of distilled, structured knowledge extracted from the medical literature and other appropriate sources.[175] (PKC referred to this repository as a "knowledge net.") It was not accessed directly by patient or clinician users. Instead, the knowledge net was a resource for building the user-level tools in (c).

b. A "coupler engine" designed for linking user-entered patient data with problem-specific knowledge in the above repository and generating output organized for decision making purposes. This engine powers the user-level tools in (c).

c. Problem-specific tools ("couplers") that employ the software engine in (b) to couple user-entered patient data with problem-specific medical knowledge extracted from the repository in (a), generating output for patient and clinician use.

Building the coupler engine in (b) is a software development project. Building the knowledge net repository in (a) and the problem-specific couplers in (c) is a long-term, labor-intensive project requiring both software developers and a team of medical content specialists trained to research, assess, and distill the medical content incorporated in the (a) and (c) components. Those components must be continuously updated over time as medical knowledge advances.[176]

---

[174]   PKC Corp. also built a problem-oriented EHR for use in conjunction with PKC's knowledge coupling tools. For discussion of actual use of both of these tools in a rural primary care practice, see the book chapter authored by Dr. Ken Bartholomew cited in Note 99. LLW intended the knowledge coupling and EHR tools to become a fully integrated set of tools, but that was never completed.

[175]   These sources now include a variety of rich datasets that were not available to PKC, including vast genomic and other data at the molecular level, and other datasets at the phenotype level. One example is the Human Phenotype Ontology, which "provides a standardized vocabulary of phenotypic abnormalities encountered in human disease." See its entry on Primary adrenal insufficiency (Addison's disease), the subject of Section 3.1.3. The HPO is produced by the Monarch Initiative, "an NIH-supported international consortium dedicated to semantic integration of biomedical and model organism data." The Monarch Initiative is part of the Global Alliance for Genomics and Health. Such organizations and their resources now make it possible to efficiently build unified CDS and EHR guidance tools that were beyond PKC's capabilities.

[176]   For more detailed discussion, see LLW's 1991 book, *Knowledge Coupling* (Note 18), especially chapter 7 (pp. 149–156. See also Shukor, A. R., An alternative paradigm for evidence-based medicine: Revisiting Lawrence Weed, MD's systems approach. *Perm. J.*, 2017;21:16–147; DOI: 10.7812/TPP/16-147.

LLW regarded the National Library of Medicine (NLM) within the National Institutes of Health (NIH) as the entity best positioned to lead creation of CDS tools implementing the combinatorial approach. See Larry Weed's Legacy and the Next Generation of Clinical Decision Support. This guest blog post observes:

> Drawing on its uniquely comprehensive electronic repository of medical content, NLM could create a new repository of distilled, structured knowledge. Drawing on its connections with the NIH research institutes and federal health agencies such as the CDC and FDA, NLM could rapidly incorporate new knowledge into that specialized repository. Outside parties and NLM itself could use that repository to build user-level tools with a unified design for conducting initial workups on specific medical problems.

> By enabling creation of such a knowledge infrastructure for the public, NLM would seize an "opportunity to modernize the conceptualization of a 'library.'" Beyond its current electronic repository, NLM could be "demonstrating how information and knowledge can best be developed, assimilated, organized, applied, and disseminated in the 21st century." [NIH Advisory Committee to the Director, NLM Working Group, Final Report, p. 12 (June 11, 2015).]

Among other benefits, such a knowledge infrastructure would dramatically improve on the status quo for enabling rapid, two-way communication regarding pandemic risks. By continuously updating web-based CDS tools for screening, diagnosis and treatment of the countless diseases for which the finding sets overlap the finding sets of infectious diseases with pandemic potential, the NLM could rapidly inform clinicians and patients of the latest actionable knowledge on all these diseases including a current pandemic or alert them to indicators of a newly identified infection posing risk of a pandemic. This emerging knowledge would automatically become apparent (not just available) to tool users at the time immediately relevant to them, helping them distinguish between similar diseases (e.g., Covid-19 and seasonal flu). Moreover, researchers could study ongoing use of CDS and EHR tools to identify, for example, emerging patterns of infection in user data entries.[177] For further discussion of these points in relation to the COVID-19 pandemic as an example, see the next section.

8.3.3   THE SEVEN-STEP SEQUENCE IN DETAIL

The process of assembling information (the first stage) is usually set in motion once a person experiences a symptom of concern or goes to a clinician for treatment of a known problem. Alterna-

---

[177] See Mercado, E., Federal gov't faces COVID-19 crisis blindly; Lacks risk monitoring system that would include pandemics. Center for Investigative Journalism, April 17, 2020. (HHS "for 12 years has failed to comply with its obligation to operate a centralized digital platform to share information, as close as possible to real time, about threats to public health such as pandemics.")

tively, a clinician may identify the problem, for example, by uncovering some abnormality during a general checkup. Regardless, the initial data point—a symptom as the person experiences it, or an abnormality as the clinician finds it—is the starting point for further inquiry.

Given the initial data point, medical knowledge about it must be obtained from somewhere. That somewhere may be online sources when the person begins the inquiry, or from the clinician's own mind when the clinician begins the inquiry. But those are the wrong knowledge sources. Neither generate carefully defined and controlled informational inputs, as discussed further below.

Regardless of the knowledge source, it's crucial not to confuse the initial known reality (the data point) with initial assumptions or beliefs about it. As an example of such confusion, suppose the patient's problem is a persistent cough. Some clinical guidance software assumes that a cough problem is caused by an upper respiratory infection (URI), which may indeed be the most common cause. The software thus provides diagnostic and treatment guidance for a URI. That assumption and guidance fail to account for the possibility that the cough is a symptom of some other cause, such as a lung tumor.[178] Another example is the endometriosis case discussed in Section 3.1.1, where four doctors prematurely accepted two different preliminary diagnoses as final, thus failing to consider and test for some other cause such as endometriosis.

To avoid pitfalls such as these, the patient and clinician must consistently follow the meticulous seven-step sequence outlined in Section 8.3.1. The following further explains the sequence in detail.

8.3.3.1 Step 1. Link the Initial Data Point (the Patient's Problem) with Applicable Knowledge

Let's return to the cough diagnosis example. The CDS tool should initially display a list of possible complaints (e,g., abdominal pain, chest pain, coughing, depressed feelings, diarrhea, fatigue, headaches, etc.), from which the user chooses the cough complaint. That stated symptom becomes the preliminary entry on the problem list. See Section 9.1.2.2. The problem list must *not* restate the problem as, say, "suspected upper respiratory infection." That suspected cause is a diagnostic hypothesis (to be stated in a care plan as an option for diagnostic investigation). Thus, it should not appear on the problem list unless and until it is confirmed. This kind of precision is required as a matter of scientific integrity and patient safety.

Having selected the CDS tool for cough diagnosis, the first step in the seven-step sequence is to link the initial data point (the cough problem) with medical knowledge about all known diagnostic options for cough, common or rare. Clinicians refer to this first step as developing a "differential diagnosis" or list of diagnostic possibilities. But clinicians vary in what their lists include.

---

[178] This example comes from a blog post by Dr. Ken Bartholomew, discussed in Section 9.1.3.2.

A complete "differential" for routine problems such as cough should include potential diagnostic associations (including but not limited to known causes) for which cough is a useful signal. For some problems, these possibilities can number more than 100. But clinicians rarely develop a complete list, especially at the outset of care when completeness matters most. Instead, "clinicians work from short lists, and these lists vary from specialty to specialty."[179] Their short lists prematurely take each clinician's thinking and observations in different directions. Those directions vary even among clinicians within the same specialty, as illustrated by the endometriosis case discussed in Section 3.1.1 and the Parkinson's disease case discussed in Section 3.1.2. Moreover, which specialty is most relevant varies from case to case and can't be determined in advance. "Patients with chronic cough," for example, "present to health care providers in a wide range of specialties, and if successful resolution is not rapidly achieved, these patients can pose diagnostic and management challenges to many clinical services and may see numerous doctors."[180] For cough, relevant specialties include pulmonology, cardiology, immunology, gastroenterology, oncology, and others. So patients can find themselves being shunted from one specialist to another for months or years.

This kind of complexity demands generally accepted standards of care for both initial diagnostic inquiry and follow-through investigation. Yet, such standards, to the extent they exist at all, are at best fragmentary, are not generally accepted, and are distorted by population-based concepts of medical knowledge.

The situation is transformed when the first step is carried out by a problem-specific CDS tool. Building the tool involves researching the *full range* of diagnostic options in *all* specialties in advance of any patient encounter. Using the tool to begin the initial patient encounter, *before* exercising clinical judgment, ensures that all options are taken into account. That completeness provides a secure basis for Step 2.

Some may object that identifying all diagnostic options at the outset is unnecessary and unproductive for sufficiently experienced, well-trained doctors, because they can zero in on the most likely possibilities. This objection evidences medicine's state of denial, as explained in detail by part IV.G.2 (pp. 85–89) of *Medicine in Denial*. See also part IV.A (pp. 53–56), discussing the special importance of "threshold processes" relative to "follow-up processes." Recall Francis Bacon's point that "the whole operation of the mind must be completely re-started, so that from the very beginning it is not left to itself ...."[181]

---

[179]   Authors' response to letters to the editor, *N. Engl. J. Med.*, 1996;334:1403–1405, about the article on the first Addison's disease case discussed in Section 3.1.3 (see Note 42).

[180]   Smith, J. and Woodcock, A., Chronic cough. *N. Engl. J. Med.*, 2016;375:1544–1551. DOI: 10.1056/NEJMcp1414215

[181]   See Note 163.

8.3.3.2   Step 2. Define Each Option as a Set of Simple Findings from Readily Available Sources

Only a portion of the diagnostic options from Step 1 will turn out to be worth considering for any individual. So the second step is to sort out which *potentially* relevant options for a cough problem are in fact *actually* relevant for a particular individual experiencing that problem. Identifying that "actually relevant" subset filters out those options that can safely be ignored for that individual. This filtering saves limited resources—time, attention, and money—from being wasted on irrelevant information.

Identifying the options of actual relevance is straightforward. Recall that the literature on the problem is researched and the CDS tool is built in advance of being used for patient encounters (as distinguished from a literature search during or after encounters). That research enables identifying all potentially relevant options and defining each one as a set of simple findings that, to the extent they are positive *in combination*, suggest the diagnostic option is worth considering for the patient. If *none* of the items in this finding set are positive (or uncertain), then the option can be safely ignored as of the time the findings are made.[182] So this step in the process requires checking all items included in the finding sets used to define all the options identified by Step 1.

For example, one of the many diagnostic options for a cough problem is COVID-19.[183] Building the CDS tool for the cough problem involves researching trusted sources to determine the findings to include in the initial workup relating to the COVID-19 option. Based on the current CDC screening tool for COVID-19,[184] symptoms commonly found if and when COVID-19 becomes symptomatic are the following:

Fever or chills	Mild or moderate difficulty breathing
New or worsening cough	Sudden loss of taste or smell
Sore throat	Vomiting or diarrhea
Unexplained, significant fatigue or aching throughout the body	

---

[182] Suppose the patient or clinician nevertheless have an intuition that the option should not be ignored, despite the absence of evidence for it. Then they can re-check the negative findings later to see if any have turned positive. The CDS tool informs them of the specific findings to monitor. So they need not rely merely on intuition.

[183] In reading what follows, bear in mind that COVID-19 is currently not representative of most conditions that doctors must diagnose. All clinicians and most other people are now hyper-aware of COVID-19 as a diagnostic possibility for some of its symptoms. That widespread awareness means that clinicians are less likely to overlook COVID-19 than other diagnoses. Even so, CDS tools would have immediate value in coping with the pandemic, because the tools would give everyone more comprehensive, current, and actionable knowledge than they have from the current widespread awareness.

[184] COVID-19 Screening Tool (as of February 15, 2021). This example illustrates how disease screening can be productive if we "carefully pick and choose" it. See Section 3.1.4, discussing Dr. Welch's critique of screening.

None of these symptoms are specific to COVID-19. On the contrary, every one of them occur with countless other conditions.[185] These symptoms become significant diagnostically for COVID-19 to the extent that some or all of them appear *in combination*.

One of the listed symptoms—new or worsening cough—is the problem to be investigated in our example (each of the other symptoms would be the subject of other CDS tools). Therefore, the CDS tool for investigating cough would pose questions to the patient about whether any of the other six listed symptoms are present.

In addition to symptoms, the CDS tool needs to address risk factors, physical exam observations, and basic lab test results. Specifically:

- **Risk factors.** Current CDC guidance states that, as to risk of "severe illness from the virus that causes COVID-19", adults of any age with 12 listed medical conditions "*are at increased risk,*" while adults of any age with 11 additional listed conditions "*might be at an increased risk*" (emphasis added).[186] Of these two lists of risk factors, the first list would reasonably be included in the finding set for COVID-19. The second list would be best included not in the finding set but in the accompanying explanation (see Step 7). Risk factors are not limited to the co-morbidities specified in the two lists. Other risk factors include advanced age, crowded living or working conditions, working in health care facility, exposure to people (or travel to locations) with COVID-19, failure to take precautions like wearing masks, and more. Data on some or all of these risk factors could be collected in Step 1, and further knowledge about them could be provided at Step 7.

- **Physical exam observations.** As yet, there do not appear to be physical exam observations (such as lung sounds, blood pressure, or heart rate) that would be useful to include in initial workups as indicators of possible COVID-19.[187] If further study reveals specific observations that would be useful to include, then those could be incorporated in the CDS tool for cough diagnosis and thus be immediately disseminated to all users of the tool.

- **Lab tests.** The diagnostic testing alternatives for COVID-19 have been developing rapidly. The PCR test to detect the viral genetic material has been the most accurate. But it would not make sense to include that test in the initial workup for everyone with a cough problem. Only a small proportion of coughing cases are caused by COVID-

---

[185] For example, coughing and breathing difficulty are included in the CDC's listing of symptoms for advanced lung cancer.

[186] CDC, People with Certain Medical Conditions, as of February 15, 2021.

[187] In addition to the CDC guidance cited in the preceding note, see Featured Review: Can symptoms and medical examination accurately diagnose COVID-19 disease?, Cochrane Infectious Diseases Group (July 7, 2020).

19, the PCR test is somewhat invasive and uncomfortable (at least in its original version), and its results may not be quickly available. So, if the output from steps 3–6 suggests that COVID-19 is plausible relative to other possibilities identified as worth considering, then the output from Step 7 could suggest the PCR test for confirmation or rule-out. But many alternatives to the PCR test have been developed, including several that can detect both coronavirus and flu viruses at the same time.[188] Such tests, if sufficiently quick, non-invasive, reliable, and inexpensive, might reasonably be included in finding set for every initial workup for cough and for other COVID-19 symptoms, at least on a temporary basis during the pandemic, even though only a small proportion of people with cough (or other COVID-19 symptoms) have the disease.

8.3.3.3 Step 3. Combine All Finding Sets for All Options from Step 2 and Generate a List of Findings to Be Made

The next step is to aggregate all the finding sets for all the options and generate a list of findings on the patient to check. As with Steps 1 and 2, the software executes this step instantaneously at the point of care. For this step, the following points are important to understand.

- Different options may share none, one, or more findings. Thus some of the options partially overlap in the findings used to define them. This happens because a single finding in isolation is typically non-specific. That overlap means the total number of findings to be made for a given problem will include duplicates, but those are removed from the aggregated list of findings to check. For that reason, and because these initial findings are quickly made with little effort for each one, the initial data collection and analysis are highly efficient.

- Nevertheless, the total number of findings to check is more than clinicians are used to at the outset of care, and certainly more time-consuming than is feasible in the typical 10–20-min encounter. This causes some clinicians (or their overseers) to reject the combinatorial approach out of hand. But this rejection is short-sighted. Using the CDS tool to guide data collection transforms the trade-off between thoroughness and speed. See *Medicine in Denial*, part IV.G.1 (pp. 80–84), discussing the feasibility of detailed data collection. With combinatorial analysis enabled by the CDS tool, true thoroughness can become a first step rather than a last resort. And thoroughness up front pays dividends far outweighing the false efficiency of snap judgments from skimpy data.

---

[188] See the FDA's FAQs on Testing for SARS-CoV-2 and Billingsley A (GoodRx), The Latest in Coronavirus (COVID-19) Testing Methods and Availability, January 25, 2021.

8.3.3.4 Step 4. Make the Findings and Record Them as Positive, Negative, or Uncertain

At this point the division of labor temporarily shifts from the CDS tool to the patient and clinician. They make the findings listed by Step 3 and enter each finding in the tool as positive, negative, or uncertain. The tool should allow the patient and/or clinician to annotate these data entries with free text. These entries of data and any free text are an opportunity for patient and clinician to consult each other and have a focused, clarifying dialogue.

For example, patients often have incomplete knowledge about some family history items, which should thus be entered as uncertain. If the patient believes she could later find out more about the family history from a relative, she could explain this in a free text annotation, which could be flagged as a reminder for later follow-up. Or if a test result is entered as uncertain because it's borderline normal/abnormal, the clinician could explain this in free text. Alternatively, if the clinician has reason to enter a borderline result as either positive or negative, the clinician could explain why in free text. (Recall the Addison's disease case in Section 3.1.3.1, where a borderline serum sodium level should have been interpreted as below normal because the patient was dehydrated.) But such free text annotation should generally not be necessary. Well designed, very specific questions posed by the CDS tool, sometimes with explanatory graphics or text, mean that a simple positive or negative response usually conveys a precise, unambiguous data point in its clinical context.

The patient should generally enter history findings before seeing the clinician. (Because the patient history is usually the most detailed portion of the initial workup, the patient rather than clinician thus assumes the largest data entry burden, possibly aided by a family member, a health coach, or staff.) Then the clinician performs the physical exam as specified by the tool, enters the observations made, orders the lab tests specified by the tool, and enters the lab results when received. The patient and clinician should discuss findings entered as "uncertain" and resolve the uncertainty if possible. The clinician should also discuss patient data entries that the clinician has reason to question, and they both should discuss physical exam and lab result entries that the patient may not understand. By thus following the tool-driven guidance and having a dialogue about the data entered, both the patient and clinician can have confidence in the completeness and accuracy of their data collection.

Then the CDS tool resumes its role, performing Steps 5–7. This arrangement is a highly productive division of labor among the patient, clinician, and the tool they jointly use.

8.3.3.5 Step 5. Match All Recorded Findings from Step 4 with All Finding Sets from Step 2

This step involves "connecting the dots"—matching all the findings on the person with the combinations of findings used to define each diagnostic option in Step 2. This matching is a simple, determinate form of pattern recognition, executable without human or artificial intelligence, "as if by machinery" in Francis Bacon's terms. How well each option matches the patient can then be expressed as "X of Y" where X is the number of positive items in the finding set and Y is the total number of items in the set. (Alternatively, users could choose to include uncertain as well as positive findings in "X" and thereby see what the effect would be if uncertain findings were in fact positive.) The output of this matching process is displayed in Step 6.

8.3.3.6 Step 6. Generate Initial Output—A List of Options, Categorized and Ranked

This step in the sequence generates output organized into groups of options and, within each group, a list of options worth considering for this patient.

The groupings for the options vary depending on the problem the CDS tool addresses. Examples of these groups include the following (for diagnostic couplers).

- **Causes that may indicate an emergency situation.** Positive findings for such diagnoses inform the user that immediate and ongoing attention may be required, without being delayed by full consideration of other options.

- **Causes for which just one finding makes consideration mandatory.** This group includes diagnoses for which a single positive finding is important enough to mandate consideration as either a cause or contributing factor, even if no other findings are positive.

- **Causes to consider on the basis of risk factors alone.** Diagnoses listed in this group have significant risk factors present. The clinical manifestations are all non-specific and are consequently less useful in the differential diagnosis. The presence of significant risk factors thus makes consideration of these diagnoses mandatory. These risk factors can be used to prioritize further investigation.

- **Other causes.** This group consists of remaining diagnoses suggested by at least one finding.

- **Causes for which critical risk factors are absent.** Causes in this group are less likely causes of the problem because critical risk factors are absent. (So these causes might be lower priority for further investigation, even if they have several positive findings other than risk factors.)

More examples of such groups with more explanation are provided in LLW's two prior books.[189]

To see the diagnostic options, the user would click on any one of the groupings and see a list of options in descending order of the number of positive findings on each option. It's crucial to understand that the ordering does not constitute a probabilistic ranking. It simply shows the

---

[189] See *Medicine in Denial* (Note 4), pp. 75–77, and *Knowledge Coupling* (Note 18), pp. 55, 330–347.

number of positive findings relative to total findings, thus providing a rough initial sense of how well the patient matches up with the various options within the grouping.

As an example, recall the case of the 15-year-old girl with Addison's disease (based on a *NEJM* article referenced in Section 3.1.3.1). The girl's chief complaint was severe fatigue. To generate the following example, we used a 2006 version of a knowledge coupling tool module (i.e., a CDS tool built by LLW's former company) for "Anxiety, Depression and Fatigue Diagnosis." With that tool, we entered (i) actual findings from the initial encounter and two more encounters within the first month described in the article and (ii) further findings (hypothetical but plausible) that the coupler's actual question sequences would have elicited (as positive, negative, or uncertain).[190] Given all the findings entered, the coupler would have generated a long list of diagnostic options (plus guidance on more needed data), divided into the following groups:[191]

- additional information to acquire [*this guidance informs the user of the need to employ additional screening, health review and wellness tools before making diagnostic or treatment decisions regarding a depression problem*];

- possible medical conditions (with depression, fatigue or anxiety as a prominent initial symptom);

- substance-related disorders;

- syndromes associated with fatigue, depression, or anxiety;

- causes for which one finding makes consideration mandatory;

- other disorders involving depression or anxiety; and

- DSM-IV-TR disorders that do not meet threshold of diagnostic criteria.

Clicking on the "Possible Medical Conditions" group displays a list of 21 diagnostic possibilities, the first 10 of which are shown here.

---

[190] The hypothetical findings entered included psychiatric symptoms, although the NEJM article makes no mention of whether the girl experienced them (it does mention, however, her doctors' unfounded suspicions of an eating disorder, poisoning by a family member, and Munchausen's syndrome). Yet psychiatric symptoms are commonly experienced by Addison's disease patients, sometimes preceding the onset of physical symptoms. If the girl did in fact experience psychiatric symptoms, then the article should have mentioned them as an overlooked clue to the correct diagnosis. If the girl did not experience those symptoms, then their absence should have been mentioned as unusual for Addison's disease and thus as possible evidence of not having that disease.

[191] This list shows only the groups with a diagnosis applicable to this patient. Numerous other groups and subgroups and explanations are omitted.

4 of 6	Adrenocortical insufficiency (for example, Addison's disease)
0 of 2	Risk factors: Adrenocortical insufficiency (for example, Addison's disease)
2 of 5	Sleep disorder
2 of 6	Pancreatic carcinoma
2 of 6	Uremia
2 of 9	Hyperthyroidism
1 of 1	Ovarian cancer
0 of 1	Risk factors: Ovarian cancer
1 of 4	Diabetes mellitus
0 of 2	Risk factors: Diabetes mellitus

The above list shows that the only medical condition matching well with the positive findings on the girl is "Adrenocortical insufficiency (for example, Addison's disease)," which was in fact the correct diagnosis as described in the *NEJM* article. But the user should not reach any conclusion based on the above information alone. More than one cause could be present. Other groups of options (besides the Possible Medical Conditions group) may include some other possible diagnosis matching the patient at least as well. And one group is "Causes for Which One Finding Makes Consideration Mandatory." Any diagnostic option in that group would need to be considered even if the Adrenocortical Insufficiency option were a perfect match with the patient and seemed acceptable as a final diagnosis. Accordingly, the user must proceed to Step 7.

8.3.3.7 Step 7. Generate Detailed Further Output on Each Option

This step allows the user to drill down to explanatory information about the options and findings in the Step 6 output. For example, clicking on the adrenocortical insufficiency diagnostic option would display a screen with the following information.

- Data for adrenocortical insufficiency—View Plan Options →   [click-through screen omitted]

- EVIDENCE for this option PRESENT for this patient

  ○ Hypotension [*References*]

    • Hypotension is a common symptom of adrenocortical insufficiency. It may be mild and limited to postural hypotension in early stages of the disorder.

  ○ Skin hyperpigmentation [*References*]

- • Increased pigmentation of the skin and mucous membranes occurs in 92% of patients with primary adrenocortical insufficiency. Hyperpigmentation appears brown, tan, or bronze on the face, elbows, knee, knuckles, palmar creases, nipples, and areolas. Hyperpigmentation appears gray on the mouth and buccal mucosa. Hyperpigmentation does not occur with secondary adrenocortical insufficiency.

  - • Numerous deeply pigmented dark-brown nevi (skin moles) may erupt on the extremities and trunk, as reported in two case reports of Addison's disease.

  - ○ Hyponatremia (if available) [*References*]

  - ○ Significant weight loss [*References*]

    - • Weight loss appears in about 97% of adrenocortical insufficiency cases. [*References*]

- • EVIDENCE for this option NOT PRESENT [or uncertain] for this patient

  - ○ Hyperkalemia (if available) [*References*]

    - • Hyperkalemia (serum potassium high) occurs in 64% of primary adrenocortical insufficiency cases [*References*]

- • Important points about EVIDENCE

  [*Under this heading are several paragraphs of discussion and references, not shown here, about psychiatric symptoms, other symptoms, and primary versus secondary adrenocortical insufficiency.*]

As another example, going back to the list of groups, clicking on the "Causes for Which One Finding Makes Consideration Mandatory" group would show only one diagnostic option, hyponatremia, which is defined by a single finding—a below normal serum sodium level. Clicking on that option would display a screen with the following information.

- • EVIDENCE for this option PRESENT for this patient

  - ○ Hyponatremia (if available) [*References*]

    - • In hyponatremia, for serum sodium levels of 122 mEq/L and above, the patient is usually asymptomatic. [*References*]

- • Important points about EVIDENCE

[Under this heading are two paragraphs of discussion and references, not shown here, about various symptoms caused by two types of hyponatremia, and about neurological dysfunction due to hyponatremia.]

8.3.3.8 Further Examining the Cough Diagnosis/COVID-19 Example

Now that we have described the seven-step sequence, let's further examine the example of COVID-19 as a diagnostic possibility for cough. Consider the following three scenarios.

- Suppose a person has recently experienced most of the seven symptoms listed under Step 2 and is thus a good match with CDC's screening profile for COVID-19, regardless of risk factors, physical exam findings or lab tests. That does not mean that the person necessarily has the disease, but it does mean that COVID-19 is a high priority for further diagnostic investigation. If one of the validated tests for COVID-19 is administered and the result is negative, then the other possible causes of the cough problem need to be considered. Moreover, other COVID-19 symptoms the patient is experiencing can still be investigated by using the CDS tools applicable to those other symptoms. This further investigation helps protect against the possibility that the COVID-19 test result was a false negative, which illustrates one way the combinatorial approach is fault-tolerant.

- Suppose all of the finding set items (symptoms and risk factors plus any physical exam items or lab tests that may later become included in the finding set items for COVID-19) are negative. That means the initial workup for the cough problem shows no evidence for COVID-19 (other than the cough itself) as a possible cause. Thus, COVID-19 is not worth considering as a diagnostic possibility and can be safely ignored as of the time the findings are made. Accordingly, COVID-19 would not be included in the output at step 6. But one or more of the symptom findings could become positive if checked at a later time, either because the disease has advanced, or because the patient has contracted the disease and become symptomatic at that later time. Whenever one of the COVID-19 symptoms appear, it should be investigated using the CDS tool applicable to that symptom. This is one way in which the combinatorial approach accounts for individual variation in how diseases present themselves over time.

- Suppose several of the finding set items are positive while several others are negative or uncertain. That means COVID-19 may or may not be a priority for further investigation, depending on the context. The context includes, for example, presence or absence of risk factors, and comparison of all diagnostic options in terms of how well their finding sets match positive findings on the patient.

8.3.3.9 Another Case of Cough Diagnosis—Curing the Problem over a Glass of Wine

As an example of carelessness and harm and waste that the above seven-step process could protect against, Ken Bartholomew (see Note 99) provides the following case. Dr. Bartholomew and one of his partners were on a hunting trip in Montana with an old friend. Over a glass of wine, the friend got talking about how he started having allergies when he moved to Montana. He found that he had a constant cough and severe shortness of breath while hunting, so he went to some "specialist," who ordered multiple tests and a CT scan. The results were normal, and he was referred to an allergist, who ordered more tests. Again the results were normal. Then steroid treatment was ordered. Months of taking the steroids led to bloating, weight gain, and facial swelling but no relief from the cough.

Dr. Bartholomew then said somewhat lightheartedly, "Just stop taking that ACE inhibitor." "How did you know I was taking an ACE inhibitor?" his friend asked. He had started it shortly after moving to Montana, when the apparent allergies first appeared. Coughing is a well-known side effect of ACE-inhibitor drugs—something that should have been known to or checked by both of the specialists that the friend had consulted months before. On the advice of Dr. Bartholomew and his partner, their friend did stop taking his ACE-inhibitor drug. Within weeks he felt "unbelievably better."

So the outcome of conventional "care" by two specialists was $20,000+ worth of testing, who-knows-how-much in doctor fees, radiation exposure from the CT scan, unnecessary expense and side effects for the steroids, months of wasted time and needle sticks and misery, and no solution for the patient's problem. By comparison, Dr. Bartholomew writes, their casual discussion "cured him over a glass of Merlot." As a further comparison, he notes that the ACE-inhibitor drug "would have come up immediately" in LLW's knowledge coupling software as a possible cause of the cough problem.

8.4 VISUALDX: A CDS TOOL OPTIMIZED FOR VISUAL INFORMATION

LLW's "knowledge coupling" CDS tools for informational guidance (the subject of this chapter) and his problem-oriented record tools for process guidance (the subject of the next chapter) are not limited to particular specialties or information types. Instead, the tools are designed for medical decision making in general. This generality becomes possible with problem orientation, which applies generic concepts of orderly problem solving in any field.

The result is simplicity—a set of tools with a consistent interface and unified design for all users in all medical contexts, even if different vendors produce specific components of the toolset. To the extent this simplicity and generality are achieved, users do not have to switch back and forth among different interfaces and designs for different specialties, different practice settings, different data types, and the like. Such generality enables intuitive ease-of-use at the user level and powerful synergies at the software development level.

But precisely because of their generality, LLW's tools are not optimized for one information type of special importance in medicine: visual information. The same is true of other general CDS tools. This issue was recognized by Dr. Art Papier. He originated VisualDx, a diagnostic and patient management system with a core design oriented toward visual information.

In medicine visual information takes various forms, clinical images among them. The brain processes visual information with pattern recognition thinking, as distinguished from analytical thinking. The former is rapid and intuitive; the latter is slower and more deliberate.[192] Both types of thinking benefit from support by external tools in the first stage (assembling information) of decision making.[193]

VisualDx aims to assist both forms of thinking for medical problem-solving purposes. Most CDS systems support analytical thinking, not the pattern recognition involved in processing visual information. Dr. Papier thus argues that VisualDx "fulfills one of the greatest areas of information need in medicine: to support pattern recognition and analytical thinking simultaneously."[194]

VisualDx is often (incorrectly) viewed as a tool for dermatology, where visual information is paramount. Dr. Papier is a dermatologist, and the early development of VisualDx focused on assisting non-dermatologists to recognize visible clues on the physical exam. The tool was deliberately named "VisualDx," not "DermDx." It now covers all general medicine illnesses, including cardiopulmonary, gastrointestinal, neurologic, renal, urologic, infectious diseases, and other areas. It integrates various types of visual information (not just images but drawings and graphics) with related non-visual information.

Dr. Papier was inspired by his exposure to LLW's knowledge coupling tools and problem-oriented record standards during medical school.[195] After completing his training in dermatology, Dr. Papier and colleagues began to develop software designed to couple visual and non-visual information (both knowledge and patient data) for decision support. The company he founded in 1999, after initially focusing on public health informatics, has become the first diagnostic CDS system to have wide impact. It is now licensed at over 2,300 hospitals and large clinics internationally, and at more than half of U.S. medical schools. No other diagnostic CDS tool has gained such acceptance.[196]

---

[192] See generally Kahneman, D., *Thinking Fast and Slow* (Farrar, Straus, and Giroux, 2011). See the Wikipedia article on that book for a brief discussion of the two forms of thinking.

[193] See *Medicine in Denial* (Note 4), pp. 112–116, where we discussed the two forms of thinking (we used different labels) as components of professional expertise. Our discussion there did not consider the role of human pattern recognition in processing visual information as an important aspect of medical practice.

[194] Personal communication with Dr. Papier. He has assisted with the drafting of this section.

[195] See the two articles about LLW by Dr. Papier cited in Note 12.

[196] Personal communication with Dr. Papier. For further background, see Warner, C., Dr., Art Papier of VisualDx: 5 Things We Must Do To Improve the US Health care System. Authority Magazine, August 9, 2020.

VisualDx leverages several aspects of diagnostic practice.

- Clinicians do not just *analyze* clinical information; as Dr. Papier says, they literally *see* (and feel and hear and even smell) their patient's condition.

- Human processing of information in visual form is often more efficient than processing the same information in textual or numerical form. That is why charts and graphs are so useful. This phenomenon is exploited in aviation, where visual displays and controls have been shown to expedite information delivery and response. VisualDx does much the same in medicine.

- Diagnosis requires processing not only visual patterns (images) but patterns of association between visual and non-visual information. Internal diseases often are manifested by external, visible clues associated with other patient data, typically from the patient history, physical exam, and lab results. Yet these associations often go un-interpreted, especially by primary care clinicians, resulting in missed or delayed diagnoses.

As an example of the last point, a patient presenting to the emergency department with both fever and a rash might be presenting with a serious, life-threatening infectious disease. Diagnosis then requires immediate, accurate recognition of the rash. Specifically, the combination of fever and a rash requires matching images of various rashes plus detailed non-visual patient data with more than 130 possible diagnoses, which include infectious, immunologic and rheumatologic conditions, environmental exposures, and more. No clinician can perform this matching function without an external tool.

Another example is hyperpigmentation such as moles or other darkening of the skin. Recall the cases in Section 3.1.3.1 (where multiple clinicians failed to recognize the patient's moles as a key diagnostic clue to Addison's disease) and Section 3.1.3.2. (where a dermatologist similarly failed to recognize the same clue to the same diagnosis but used an Internet search for causes of hyperpigmentation to quickly identify Addison's as a diagnostic possibility).

These examples suggest that non-experts, particularly generalists, have only limited ability to process visual information and associated patterns, because they have not seen a wide enough range of patterns, and their ability to memorize and recall these patterns is limited. Augmenting the human brain with the full-spectrum of possible diagnostic patterns becomes possible through visually-based decision support. New forms of decision support will take advantage of machine learning and AI to assist in the recognition of patterns, particularly for generalists. VisualDx, by incorporating a problem-oriented knowledge base with machine learning of imagery and graphical representation of the diagnostic possibilities, helps make this enormous complexity manageable.[197]

In short, VisualDx optimizes a broad range of visual information and integrates the multiple ways it can be used diagnostically. These ways include viewing external tissues (and photos thereof), imaging of internal tissues by various technologies, graphically representing numerical and temporal patterns, using medical illustrations to highlight information, and combining visual and other information by use of traditional software and artificial intelligence.

The following outlines some of the specific ways VisualDx optimizes a visual approach to diagnosis.

1. Dr. Papier and his team have found that "that medical imagery and related case data require the same sort of editorial control and quality as written medical publications." Accordingly, the company has spent 20 years accumulating the world's largest medical image collection with associated de-identified case data. The images are highly curated, including restoration of deteriorated images as part of digitization. The company describes its image collection as follows:

   > Our goal with images is to represent the full spectrum of human disease across skin types, body location, morphologic variation, and other diagnostic features. To adequately capture the immense variation inherent in disease requires access to hundreds of specialized image collections that span dermatology, ophthalmology, oral medicine, infectious disease, emergency medicine, radiology, and pathology. For this reason, we've established partnerships with many exceptional learning institutions and individuals who have contributed their images toward this effort.

2. VisualDx captures the reality that a single disease may be manifested in enormously variable ways, with the variations occurring among different patients or within a single patient over time. Thus, for example, VisualDx provides multiple images for a particular type of skin lesion, showing how the same lesion type may vary in appearance (for example, depending on race—see #6). Such variations can easily go unrecognized, leading to missed diagnoses. See Chapter 7 on page 96, discussing how these variations are obscured by the fallacy of the "classic" or "textbook" case, and the misleading generalizations of medical knowledge. LLW's knowledge coupling tools also highlighted unique individual variations, but did so largely without image-based presentations of these variations.

---

[197] Some CDS tools make limited use of visual information. For example, in LLW's knowledge coupling tools, patient history and physical exam questions often use illustrations for indicating the location of pain or showing physical exam elements. In addition, LLW long emphasized the importance of medical record flowsheets to represent some data types in tabular or graphical form. See "Medical Records That Guide and Teach" (Note 188), PDF pp. 8–9. But such uses of visual information fall far short of what is possible.

3. VisualDx enables clinicians to rapidly input patient data and receive back a graphical and image-based representation of diagnostic possibilities and how they match with the data. For example, the "Sympticon" feature lets clinicians visualize and compare symptoms by a series of searchable graphics. Each Sympticon highlights which organs are affected by the specific disease or diagnosis. For example, an osteoarthritis Sympticon is an image showing locations in the body where each of the following symptoms appear: limited range of motion, polyarthralgia, bone pain, joint stiffness, low back pain, hip pain, knee pain. The image of this pattern of locations facilitates matching the patient's pain symptoms with the diagnostic possibility of osteoarthritis.

4. Medication side effects and interactions commonly cause adverse events, such as visible skin lesions. VisualDx includes a comprehensive database of medications that maps to conditions and adverse events triggered by medications. This feature helps clinicians take into account medication effects as diagnostic possibilities. In addition, searching the database by medication provides images, evidence, and therapy guidelines.

5. VisualDx includes an AI machine learning feature, which is combined with rules-based decision support. This feature analyzes a photo of a skin lesion, guides the clinician on basic data collection about the lesion, and generates a list of diagnostic possibilities. [198]

6. As discussed in Section 3.3.4.4, a major issue with AI tools is the potential for discrimination arising from biased datasets used to train the tools. A prime example is under-representation of non-white populations in dermatology image datasets. The effect is to compromise accurate diagnosis. The same skin lesion type caused by the same diagnosis may have different appearances in light and dark skins. So an AI tool might not recognize the correct diagnosis. The same discrimination may occur with human intelligence and traditional software, both of which rely on dermatology image libraries. An example is melanoma (a skin cancer, fatal if left untreated), which is often missed or misdiagnosed in Black patients.[199] Moreover, "when people of color have their clinicians explain a disease process with images that do not look like them or their condition, it creates confusion, undermines trust, and ultimately affects patient behavior and outcomes."[200] Fixing failings like these requires long-term, con-

---

[198] See DermExpert FAQs, which includes some technical specifics on VisualDx's AI/machine learning capability.

[199] Mahendraraj, K, et al., Malignant melanoma in African–Americans. *Medicine* 2017;96:15. DOI: 10.1097/MD.0000000000006258.

[200] See Papier, A., To begin addressing racial bias in medicine, start with the skin. *STAT First Opinion*, July 20, 2020.

certed efforts, including creation of better image datasets, in which VisualDx has had a leading role for years.[201]

7. VisualDx incorporates a function called "guided work-ups." Following the exemplar of LLW's knowledge coupling CDS tools, VisualDx provides its users with problem-oriented questionnaires (the contents of which depend on the problem—the patient's chief complaint). VisualDx embraces LLW's principle that busy clinicians typically cannot know all the specific data points to gather from the history, physical exam and lab tests. Instead, as LLW was among the first to recognize, more comprehensive questionnaires can be completed by patients at home or by para-professionals. In creating the knowledge base for this data collection, VisualDx created a concept-oriented terminology and data structure to support history taking in patients' language. The patient's data entry at home could thus be used by the clinician because of this concept-oriented terminology. In this way VisualDx is enabling two-way information exchange between patients and clinicians, a system it calls "Collaborative CDS" or "Co-CDS."

In summary, VisualDx is a problem-oriented CDS system derivative from LLW's ideas and tools for knowledge coupling CDS. It seeks to balance the heuristics of problem-solving by augmenting pattern recognition when needed, while offering a pathway for more deliberative analytical thinking. Not all problem solving in medicine follows the same thinking and problem-solving brain pathways. In some instances, immediate diagnosis can be achieved from a single, instantaneous visual clue. A skilled diagnostician, or now a generalist assisted with augmented AI-based machine learning, can see a particular lesion or rash and "know" the diagnosis with 99% certainty instantaneously. On the other hand, a definitive diagnosis often cannot be immediately discoverable, and diagnosis is an analytic process where the "final" diagnosis unfolds over time.

At the onset of the diagnostic process is the need to have a contextual, full data set. Complex and ambiguous diagnostic scenarios require information tools, so the right questions are asked and patients are examined in the context of their complaint. Through CDS we can assist the brain so both patients and clinicians ask the right questions. VisualDx provides this assistance with a form of guidance not offered by CDS like LLW's knowledge coupling tools. VisualDx uses iconography and, where possible, machine learning analysis of imagery, to augment the data inputted. VisualDx also differs from LLW's approach by elevating pattern recognition in the interface, supporting end user cognition by augmenting thinking through graphical representations and imagery.

---

[201] See New in VisualDx: Impact of Skin Color on Clinical Presentation, and More (December 21, 2020). See also The Impact of Skin Color and Ethnicity on Clinical Diagnosis and Research (4-part conference series, October 28–December 2, 2020), and this VisualDx website page, An Ongoing Commitment to Equity in Medicine.

CHAPTER 9

Process Guidance:
The Problem-Oriented Record

9.1 PROBLEM-ORIENTED DESIGN FOR MEDICAL RECORDS

Having described knowledge coupling tools for informational guidance in the preceding chapter, now we turn to medical record tools for process guidance.

Traditionally, medical records were passive repositories for clinicians to store data about patients' medical conditions and the care provided. These repositories merely compiled factual data. Because they were source-oriented (see Section 9.1.2.1), they did not clearly or consistently record the logic and purpose of actions taken on each patient problem.

In contrast, LLW conceived the medical record as more than a passive data repository. The record should be actively used, he argued, as an information tool for guiding the processes of care.[202] LLW saw patients, not just clinicians, as users of the tool and needing its guidance. Guidance includes communication among the clinicians involved plus the patient—each one of them must be guided by awareness of what the others are doing. Communication with the patient is fundamental. The National Academy of Medicine recognized as much when it defined diagnostic error as including failure to communicate the diagnosis to the patient.[203]

To serve as a process guidance tool, LLW argued, the health record must be problem-oriented—that is, the record's structure and contents must be oriented toward defined patient needs (problems), not provider or payer expectations. Section 7.2 discusses the meaning of problem orientation in detail.

Problem orientation as applied to an EHR system involves three core requirements.

---

[202] See Weed, L., Medical records that guide and teach. *N. Eng. J. Med.* 1968;278:593-600, 652–657 (parts 1 and 2, available here).

[203] *Improving Diagnosis in Health Care* (Note 18) , p. 4:

The committee's definition of diagnostic error is the failure to (a) establish an accurate and timely explanation of the patient's health problem(s) or (b) *communicate that explanation to the patient.* The definition employs a patient-centered perspective because patients bear the ultimate risk of harm from diagnostic errors. …The inclusion of communication is distinct from previous definitions, in recognition that communication is a key responsibility throughout the diagnostic process. *From a patient's perspective, an accurate and timely explanation of the health problem is meaningless unless this information reaches the patient so that a patient and health care professionals can act on the explanation.* [Emphasis added.]

1. **Completeness:** The record should identify the full range of patient problems requiring attention, not just the problems relevant to current provider institutions or third-party payers. (Satisfying this requirement ultimately demands consolidating records from different providers into a single record for each patient, as discussed in Section 9.2.)

2. **Structure:** Once problems are identified from initial screening, subsequent data entries should be grouped by the patient problem(s) to which they relate, not by provider sources generating the data. Within each problem grouping, data should be organized into structured categories,[204] as distinguished from unstructured aggregations of data and impressionistic narrative notes. Simply by populating these categories with data, users organize both their actions and records of care, thus enhancing both primary and secondary uses of the record.

3. **Verifiability:** Problems should be stated in verifiable terms, and the data categories should be structured to guide the verification process for each problem. In this way, the record becomes oriented toward known realities presented by unique patients, not clinicians' unverified beliefs and not population-based generalizations.

When these requirements are satisfied, the patient record fosters patient engagement, informed decision making, effective communication and coordination of care, scientifically rigorous clinical research, continuous quality improvement, and accountability for everyone involved. These goals are not achieved by merely creating interoperability among multiple record repositories or aggregating those into a single record (see Section 9.2).

9.1.1 COMPLETENESS

Most patients have a variety of health-related needs. These include not just the "chief complaint" that brings the patient to the clinician's office, and not just the medical problems within clinicians' specialty interests. Indeed, a patient's health needs encompass not just "medical" problems but also health maintenance, psychosocial problems, and problems driven by social determinants such as poor housing or isolation.

A complete problem list enumerates all the patient's health-related problems. Creating and maintaining the list not only protects against overlooking problems but also provides the total context for solving each problem. This latter point is crucial, because sound decision making on any one problem frequently requires taking other problems into account. Thus, a problem list does not (as some have objected) artificially compartmentalize the "whole patient" into separate problems.

---

[204] Compare the term "structured data," which is often used to mean data entered with a coding scheme such as SNOMED or ICD-10. This book uses the structure concept differently to mean a hierarchical structure of relationships among defined categories of data. In this sense, a record composed entirely of coded data would nevertheless be unstructured if the data are not organized into the defined categories.

Rather, the problem list is a tool for handling each problem in light of the others and thereby taking the whole patient into account.

In short, building a complete problem list should be treated as a standard of care that health professionals and institutions have a duty to follow. Without a tool and standard of care for systematically considering an explicit list of all problems, clinicians tend to take into account only the problems in their own lanes—the opposite of patient-centered care.

The goal of a complete problem list raises the question of what is meant by completeness. It does not mean exhaustiveness. Rather, completeness is a function of the data collected. A problem list is complete when it accounts for all health concerns identified from (a) screening data, (b) further data inputs by the patient and clinician raising concerns not elicited from the screening data, and (c) additional data collected to investigate the concerns in (a) and (b).

The goal of a complete problem list also raises the questions of what constitutes a "problem" for inclusion on the problem list, and who is responsible for maintaining the list. A full discussion of those and related questions is beyond the scope of this book.[205] Suffice it to say that the problem list should not be a grab bag of useful data points included on the list merely for ease of reference. Instead, the problem list should identify health issues that each require a set of related problem-solving activities over time. As to maintaining the problem list (and the record as a whole), that responsibility might well be appropriate for a health coach working closely with the patient and provider personnel as needed.

9.1.2 STRUCTURE AND DATA CATEGORIES

9.1.2.1 Top-Level Structure

It follows from the above that two initial components of the POHR structure are an initial screening database and an initial problem list derived from that database. But then, as care is provided over time, the question arises of how to organize further data accumulating in the record.

LLW observed that data organization is usually source-oriented—that is, data are grouped by their source from hospital/clinic functional areas. Source-oriented records thus typically included sections for doctor notes, nursing notes, lab data, imaging reports, procedures, vital signs, medications, specialty consultant reports, and the like.

But a source-oriented grouping or view is rarely meaningful for clinical purposes. Source orientation means that the data points relevant to any one problem are scattered. Such disarray makes it difficult to know what is going on with each problem. Equally difficult is knowing the "why" of actions taken by clinicians. Even if they record their reasoning, that information and related data

---

[205] For further discussion, see *Clinical Problem Lists in the Electronic Health Record* (Adam Wright, ed., Apple Academic Press, 2014). This book is a valuable collection of articles.

need to be seen in a problem-oriented view of all the entries relating to the problem(s) to which the action relates. For example, a source-oriented view would show an order for propranolol but not clearly indicate whether this drug was ordered for the patient's hypertension problem or her migraine problem or both (unless that order stated the purpose, which typically it would not).

The POHR structure, once the problem list is initially formulated, means that subsequent data points are entered under the problem(s) to which they relate (the problem list thus becomes a table of contents). Simply grouping data under the relevant problem in itself makes a problem-oriented view more transparent, informative, and usable than a source-oriented view of the same data.[206]

Patient records include data about both the patient's condition and problem-solving activities by multiple parties (including the patient/family). Problem-orientation thus facilitates coordinated care by implicitly organizing those multiple parties into functional teams. That is, everyone working on a given problem will enter the data they generate under that problem. Suppose, for example, the problem begins as hip pain, and a decision is made to do a hip replacement. Ultimately, the practitioners involved might include a primary care doctor, a physical therapist, an orthopedic surgeon who specializes in hip replacements, a hospitalist and hospital nurses, pharmacists, and, after the patient's return to work, an in-house clinician at a factory worksite. Each of them and the patient can thus readily ascertain what has been happening with the hip problem.

To summarize the discussion so far, the top-level POHR structure is:

• initial database, not oriented to any particular problem;

• problem list; and

---

[206] Most EHRs offer a problem list feature. "However, collection of clinical data in a logical problem-oriented format has been minimally realized," as stated in the recent article, "Impact of a problem-oriented view on clinical data retrieval," cited in Note 125. EHRs typically permit clinicians to enter data without linkage to a problem. The outcome is that data entries in many EHRs lack a problem-oriented structure. The cited article describes a project to impute a problem-oriented view (POV) after the fact. It does so by means of EHR "data summarization" in a problem-oriented manner. The project involves creating a "problem-concept map" (PCM) that links a cluster of related problem codes with codes for medications and lab results relevant to the problem cluster. The mapping of medications and lab results to problems is based on a national expert consensus. This mapping is used to infer linkages between problems and drug/lab entries in actual EHRs. (Links for data from imaging, procedures, clinic notes, and hospitalizations will be included later.) "When active in a patient chart, the POV leverages the data codes within the PCM knowledge base to retrieve relevant patient data and create a detailed display for each problem of interest." The POV thus "provides an on-demand focused view of relevant information for a given problem." Note that the POV is inferred after-the-fact by applying the PCM to an existing EHR. The inferred linkage may or may not be what the clinician had in mind when entering the data. This approach differs from LLW's conception. He intended clinicians to think carefully about what problems their actions address and to record that thinking by entering their notes and other data under the relevant problem on the problem list. LLW's approach thus builds the POV into the original data entry-problem linkages, and eliminates the need for on-demand inference and data summarization after the fact.

- problem-oriented data (problem-specific entries linked to applicable problem(s)).

9.1.2.2 Data Categories

Below the top-level structure, the POHR standard requires a substructure to organize the processes of care over time. This requires clinically functional data categories, as outlined below.

Those categories correspond to questions that the best clinicians try to ask. But many of those questions may never be asked in practice. It is difficult to think through what to ask under real-world time pressures. Moreover, even when clinicians do ask the right questions, the answers are too often hard to find (or missing) from disorganized records. The outcome is that frustrated clinicians neglect important questions just to get through the day.

In contrast, if the structure of data categories outlined below is built into the EHR user interface, the care team is prompted to populate the categories. Doing so leads them to follow high standards of care for orderly problem solving over time.

At first glance, the structured categories listed below may seem unduly complex. In practice, however, the structure is simple and intuitive. It follows common-sense steps for orderly decision making in any field: gathering information, defining problems, formulating plans for each problem, executing the plans, gathering feedback, and making corrective adjustments as needed. Organizing an EHR around these familiar steps promotes ease of use and shared understanding for every stakeholder—which current EHRs notoriously fail to achieve.

The structure of data categories[207] is as follows.

- **Initial[208] database** (not problem-specific)

  ○ *Screening data* (detailed history, physical exam, and basic lab data; allergies and intolerances; SDoH, genetic, environmental and workplace factors), all selected in advance to identify health problems and risks (this data collection should include review of existing medical records, if available)

---

[207] Those data categories should be compared with the categories in the United States Core Data for Interoper¬ability (USCDI) as modified by the ONC final rule under the 21st Century Cures Act. (On January 12, 2021, ONC released Draft USCDI Version 2 for public comment.) The USCDI categories differ considerably from what is outlined below. (The data categories should also be compared with relevant FHIR resources, as indicated in several footnotes below.)

[208] "Initial" is to be distinguished from "follow-up." The categories of initial database, initial investigation, initial problem definition, and initial care planning are boldface in the following outline, because they are disproportionately important. So too is the problem list, which has preliminary, initial, and updated versions. See *Medicine in Denial*, part V.A, pp. 53–56, discussing the crucial role of "threshold processes" relative to "follow-up processes."

- *Patient profile* (narrative description of typical day; health concerns articulated by patient not otherwise included; positive attributes relevant to decision making such as good physical conditioning)

- *Vital signs* (recorded during each patient encounter at a minimum)

- *Non-clinical data* (demographics, provider IDs, insurance coverage, other administrative items)

- **Problem list** (a preliminary version based on the Initial database, then a revised version after the Initial investigations and definitions of each problem on the list, then later updated versions). This problem list is divided into three groups:

  - health maintenance problems,

  - active problems, and

  - inactive (dormant or resolved) and transient/non-recurrent problems.

  A code from a designated coding system[209] such as SNOMED could be preliminarily assigned to each problem at this stage, but the coding should be reconsidered after the next two steps (initial investigation and problem definition). At that point, each problem will be better defined. Then some separate problems defined preliminarily as undiagnosed symptoms or abnormalities might be consolidated into one problem defined at a higher level of diagnostic understanding (a pathophysiologic condition or confirmed diagnosis).

- **Initial investigation** (problem-specific history, physical exam, lab tests, and SDoH data collection) for each problem (other than inactive/temporary ones), with entries from output of the problem-specific CDS tools as described in Section 8.3.

- **Initial problem definition** Once the user clicks on a problem entry in the list, the following elements[210] for defining the problem in further detail should be displayed:

---

[209] Sometimes the coding system might lack a code/name precisely fitting the problem as identified by the clinician. In such cases, the clinician could use free text to name the problem and also assign a code that comes closest to the free text name. Then researchers may later analyze all free text problem list entries associated with that code, and refine the code if warranted by the analysis.

[210] In addition to these elements, some believe that the record should include a narrative of "the patient's story" of how the problem emerged and/or how it affects the patient's life. Including such a narrative could add value even if the narrative details have no particular clinical significance. The clinician's mind may connect the patient's story to clinical specifics, making the problem more vivid and memorable to the clinician and also helping the patient feel his or her concerns have been recognized.

○ *Goal*: For example, the goal for a hypertension problem might be to get the diastolic below a specified level and determine the cause. But goals must be set in light of the complete problem list and life situation. If the patient has terminal cancer, for example, the goal for the hypertension problem might be to ignore it.

○ *Basis*: If the problem is a confirmed diagnosis, then the basis would be the confirmatory findings (which might be a definitive test result, or a cluster of findings established in the literature as a sufficient basis for the diagnosis). Otherwise, the basis could be a pathophysiologic condition, or an undiagnosed finding (such as a symptom, abnormal physical exam finding or test result) or other concern (such as a risk factor, family history item, or SDoH item) important enough to require ongoing attention as a separate problem. A confirmed diagnosis (e.g., cirrhosis) might cause a condition (e.g., ascites) that itself requires separate management, in which case the ascites would be the basis of a separate problem labeled as secondary to the cirrhosis problem.

○ *Status*: Statement of current progress or lack of it, indicating generally that the problem is getting better, getting worse, or staying the same.

○ *Disability*: Nature of the impairment resulting from the problem in light of the patient's lifestyle, occupation, and other functional needs.

Because the preceding four categories may change over time (as the problem evolves and as diagnostic and treatment processes advance), they can be regarded as part of ongoing progress notes separate from the problem list. But they should also be displayed up front when opening up a problem in the list, because they define the current nature of the problem.

- **Progress notes** in the SOAP note format[211] recording the categories of data listed below. "Progress notes should be written with the original [care] plan clearly in mind," rather than being mere "daily observations." This approach reduces the clinician's burden. "There is not time in a busy practice to think in depth about every problem at every encounter. Without a good plan and conscientious efforts to follow it, progress notes can degenerate into random recordings of whatever parameter strikes the fancy of a given observer at the moment." So there should be "a real effort to measure progress against established goals with well-defined parameters."[212] Moreover, notes should be conceived and entered in relation to a particular problem. Doing so provides context

[211] As an alternative to SOAP notes, some have advocated an "APSO" format where the assessment and plan come first. See LLW's critique of the APSO format and this post on the subject. See also the first bullet below.

[212] *Knowledge Coupling* (Note 18), p. 116.

and disciplines thinking. Entering a single progress note for an entire patient encounter is thus incorrect unless the encounter addressed only a single problem. Because problems may be interrelated, a note could relate to several problems. So the EHR system should allow one-time entry of text common to those several problems and then allow editing the note entry under each problem to clarify problem-specific aspects.[213] In no event should bulk cutting and pasting of prior notes be treated as acceptable. This common practice is notorious for making records bloated and hard to use.

- *Subjective*[214]*/Symptomatic data* from the patient. Of the SOAP note components, this one should be entered first. By doing so, the clinician "ensures that progress is assessed from the patient's point of view," which is the essence of patient-centered care. Doctors have a tendency to focus unduly on their own assessments of "objective" data. So they need "to keep reminding themselves that patients come to have their symptoms explained and relieved—not just to have their X-rays reported as normal."[215] Moreover, the patient's personal reactions provide important clinical feedback when they do not parallel objective data (the discrepancy might suggest a data error or misstatement of the problem). See this report of diagnostic failure illustrating "the necessity for doctors to keep looking and not brush off what patients tell them."[216]

- *Objective/Other data* such as physical exam findings, nursing data entry, test results, radiologist reports, all in either narrative or tabular form as appropriate to the data type (patient-generated data from patients' personal devices could be included here). Much objective data is best recorded in tabular flowsheets rather than text. Clinicians "are frequently inclined to omit physical findings and emphasize laboratory data. For example, there may be a detailed discussion of an x-ray and no mention of how the patient coughs or of the quantity and character of the

---

[213] "The physician should resist the temptation to write a single progress note and then say it covers four problems such as renal failure, congestive failure, chronic obstructive lung disease, and hypertension merely because they are interrelated and it seems inefficient to deal with each problem separately. The sets of parameters required to objectively assess the progress of each of those problems are not identical; and if each is not focused upon in an organized manner, crucial elements on one or more of the problems are frequently lost sight of and the whole exercise may even degenerate into general phrases like 'doing well' or 'situation not improving.'" *Knowledge Coupling* (Note 18), p. 115.

[214] The terms "subjective" and "objective" have been criticized as biased against patients and contrary to LLW's original intent to prioritize the patient's perspective. Better terminology might be "symptomatic" and "other" (these terms being compatible with the familiar SOAP acronym, as discussed by *Medicine in Denial*, pp. 167–68.

[215] *Knowledge Coupling* (Note 18), p. 116.

[216] His anemia was followed by searing foot pain. Seventeen years later, in precarious shape, a stellar athlete learned what was wrong. *Washington Post*, June 27, 2020. The same point about not brushing off what patients say is illustrated by the endometriosis and Parkinson's disease cases in Sections 3.1.1 and 3.1.2.

sputum."[217] That deficiency should be a basis for scrutinizing these record entries and improving clinician performance.

○ *Assessments*:[218] This category (NOT the problem list) is where the clinician should identify diagnostic hypotheses and treatment options to consider, which then require actions as specified in the Plans category below. Arguably, this need not be a separate data category, because assessments can be usefully stated in care plans (see the next bullet). For that reason, LLW did not include "assessment" as a separate progress note component in the computer-based POHR described by Ken Bartholomew.[219] But because clinicians are so familiar with the "assessment" component of SOAP notes, it may be worth keeping, as an alternative for entering assessment data.

○ *Plans*[220] (**initial plans** and follow-up plans) for diagnosis, treatment and monitoring, as follows (each of these categories could include explanatory assessments):

   • *Follow course—Sx and Rx*: The initial plan would specify parameters to monitor (and possibly therapeutic actions to manage difficulties the patient might be experiencing). The data resulting from this plan would be recorded in later SOAP notes under the Subjective/Symptomatic category.

   • *Follow course—Obj and Rx*: Same as preceding category, but focused on "objective" parameters. The new data resulting from this plan would be recorded in later SOAP notes under the Objective/Other category.

   • *Investigate further*: Follow-up plans for whatever further decision making is needed over time. For example, a "rule-out" plan might consist of a test to rule out a diagnostic possibility. Clinicians sometimes include a "rule-out diagnosis" on the problem list. Doing so is a serious error, conceptually and practically. Ruling out is a plan, and the diagnosis to be ruled out is *not* an identified problem. If that diagnosis is indeed ruled out, it will come off the problem list. But then the still-undiagnosed problem can easily fall through the cracks and be neglected.

---

[217] *Knowledge Coupling* (Note 18), p. 117.

[218] Compare the FHIR Resource ClinicalImpression. For further discussion of this category, see Putting the "A" Back in SOAP Notes: Time to Tackle An Epic Problem (Bob Wachter, 9/03/12), and the numerous comments, including a detailed 9/10/12 comment from LLW. Although provider records do not traditionally include patient assessments of their own problem, such patient input would fit into this category.

[219] See Note 99.

[220] Compare the FHIR Resource CarePlan.

- *Complications to watch for*: Anticipated possible complications that the care team and patient should be alert for (e.g., drug side effects, post-surgery pain or infection, interactions with other problems), and specified measures to identify and cope with complications as soon as they appear.

Progress notes in general and plans in particular are usually iterative, in that the care team is constantly circling back to enter new SOAP notes, including new plans, as prior plans are executed, feedback is received and problems evolve. Of course, if a cure or complete remission is achieved, then the problem is moved from the active to the inactive problem list.

The above discussion of data categories is not complete. Other categories found in health records are referrals to consulting specialists, reports issued by those consultants, and hospital discharge summaries. Each of these should be problem-oriented. For example, a referral should be recorded under the problem that the referring doctor wants the specialist to address, and the referral should note other problems that may need to be taken into account. Similarly, hospital discharge summaries should be problem-oriented, systematically addressing each problem for which attention may be needed after discharge. This practice is important for avoiding unnecessary readmissions, among other reasons.

9.1.3 VERIFIABILITY

9.1.3.1 Scientific Integrity

Patient records and the practices they document may fail to satisfy the verifiability requirement in various ways. One way is using an unverified diagnostic hypothesis, in lieu of the undiagnosed presenting problem, to name/code a problem on the problem list. This is a violation of scientific integrity. No scientist would record hypothesized conclusions (in lieu of observed findings) as data. Doing so would constitute research misconduct, defined by the National Science Foundation to "include changing or omitting data or results such that the research is *not accurately represented in the research record*."[221] Yet, something not unlike this and other forms of misconduct has long been endemic in health records, and increasingly in electronic records.

Dr. Mark Aronson, a former student of LLW's, has written of current electronic records:

… now it has become difficult to believe that everything documented in medical records was actually done. When you read in a chart that "A 12-point review of systems was entirely negative," do you know what was asked of the patient? When you read

---

[221] 45 CFR 689.1(a)(2) (emphasis added).

records with long and detailed templated reviews of systems and physical examinations, do you really trust them?

> Copying and pasting are ubiquitous now … . But when you copy and paste in information, have you reviewed all the data? Do you know they are accurate? Are they relevant to the problems being addressed?[222]

Medical decision making is essentially a scientific research project, especially for patients with chronic disease and other complex conditions. Medical recordkeeping should thus be conceived in terms of scientific rigor and integrity; poor records should be seen as bad science or outright scientific misconduct.

Research is routinely needed for decision making because of the imperfect fit between unique patients and the population-based generalizations of medical knowledge. See *Medicine in Denial*, pp. 51–52, 177–94. So the clinician's thinking and recordkeeping must honestly reflect patient realities, without distortion by medical "knowledge," clinician preconceptions, and financial interests of providers or third-party payers. For example, as Dr. Aronson wrote of his experience with LLW:

> Weed reasoned that forcing oneself to identify problems only to the level one understood them (e.g., dyspnea rather than chronic obstructive pulmonary disease or heart failure) helped *keep a doctor's mind open to other possibilities*. Dyspnea is, after all, complex. (Could the dyspnea be interstitial or restrictive lung disease? Could it be deconditioning? Is the patient anemic?) Writing plans so as to approach diagnoses by taking time to consider the differential diagnosis, monitor medication side effects, and order medicines in a logical and orderly manner, would not only improve patient care but would, Weed believed, also *help doctors understand the nature of their patients' diseases more completely*. He thought, and further taught, that the medical record should guide and teach. And that if it was done well, it could be audited to determine if the care was thorough, reliable, analytically sound, and carried out in an efficient manner.[223] [Emphasis added.]

LLW started developing these principles 30 years before he began work on knowledge coupling tools. Those tools not only "keep a doctor's mind open to other possibilities" but raise those

---

[222] Aronson, M., *The purpose of the medical record: why lawrence weed still matters. Am. J. Med.* 2019;132(11):1256-1257. DOI: 10.1016/j.amjmed.2019.03.051. See also Berdahl, C. Concordance between electronic clinical documentation and physicians' observed behavior. *JAMA Netw. Open.* 2019;2(9):e1911390. DOI: 10.1001/jamanetworkopen.2019.11390. This study of emergency medicine residents found that only 38.5% of "review of systems" documented in electronic records were confirmed by audio recording data, and only 53.2% of physical exams were confirmed by concurrent observation. This may reflect widespread use of auto-populated text.

[223] Aronson, M. (see the preceding note).

possibilities without any dependence on the doctor's mind. In an environment where CDS tools play that role, recordkeeping is no less important.

In short, scientific integrity in medical records is foundational to medicine, just as the integrity of accounting records is foundational to the domain of commerce (see Section 6.2).

9.1.3.2 Example: Cough Diagnosis

When the record fails to clearly distinguish among an undiagnosed finding, a diagnostic hypothesis, and a confirmed diagnosis, the resulting confusion is a recipe for medical error (not to mention fraudulent upcoding). Dr. Ken Bartholomew provides an example in his blog post, Electronic Medical Records 2017: Science Ignored, Opportunity Lost:

> Many of the current click-point programs set up a progress note labeled by a diagnosis when the patient presents. An example is the "Upper Respiratory Infection" note that the nurse will start when she rooms the patient with a cough. Questions about an upper respiratory infection will be asked, heart and lungs and ENT exam findings clicked in, and then the provider will diagnose either a viral or a bacterial URI and maybe prescribe an antibiotic. However, the early lung tumor will not be diagnosed because the "diagnosis" of URI was actually used as the presenting finding instead of the "problem" the patient actually presented with, which was cough.

The POHR standard of care protects against this corruption of the diagnostic process.

Consider how the data categories outlined above would apply to Dr. Bartholomew's example. Working with the patient, the clinician must populate the following data categories.

- **Problem list entries, preliminary**: Cough recorded as the presenting problem.

- **Initial investigation**: This investigation must methodically work through all possible diagnoses for cough problems. That can be accomplished with thoroughness, reliability and efficiency only by employing a CDS tool for cough diagnosis, as described in Section 8.3.3.

- **Initial problem definition/Basis** (i.e., results of the Initial investigation): Stating the cough finding as the problem, if the Initial investigation does not establish a clear basis for diagnosing one of the relevant possibilities and ruling out alternatives. If one of those possibilities can be established as the final diagnosis, then the problem list entry should be revised to state that diagnosis, and the "Basis" category should specify the findings justifying that diagnosis. Sometimes the Initial investigation findings will make it possible to restate the presenting problem as caused by a pathophysiologic abnormality. Then one should ask, "what caused the cause," i.e., what are the potential causes of that abnormality, in order to reach a final diagnosis.[224]

- **Progress notes/Assessment**: This category should record initial conclusions as to which diagnostic possibilities are worth considering for this patient (which might include bacterial URI, viral URI, COVID-19, pneumonia, acid reflux, asthma, lung tumor, and many others). The assessment should indicate which possibilities are priorities for testing and other investigation.

- **Progress notes/Initial Plans, specifying further data to collect**. E.g., confirmatory or rule-out tests, parameters to follow, treatments to initiate, and other actions to implement the Assessment.

- **Progress notes/SOAP entries**, recording follow-up results (symptomatic and other data) produced from the Initial Plans and follow-up assessment with further plans

- **Progress notes/Plans/Investigate further,** recording follow-up plans for whatever further decision making may be needed.

The above structure of data categories defines a pathway or process for orienting care toward the actual problem as it exists in the unique patient. That actual problem typically differs from the abstract problem as conceived in the clinician's mind or the medical literature. Thus, for example, the record must distinguish between a diagnostic hypothesis and a confirmed diagnosis, with the latter based on actual findings made and associated medical knowledge about what the combined findings mean.

Now consider what diagnosing a cough symptom involves. Scores of possible causes must be taken into account. For each cause, several findings must be made to evaluate whether that cause is worth considering in that patient. This adds up to hundreds of findings, which then must be analyzed in light of vast medical knowledge. Such thoroughness should be a first step (the initial workup), but in practice it is a last resort (or it never happens), even though such detailed findings at the outset of care are entirely feasible with the right guidance tools. Lack of thoroughness at the outset of care can easily lead to endless, futile trial and error—as the cases in Chapter 3 show.

---

[224] The question "what caused the cause" remains even after a final diagnosis is reached. But at that point, the question of causation is more relevant to public health officials than clinicians. (This illustrates how causation is a matter of purpose and context.)

9.1.3.3 Example: Opioid Treatment

As discussed below, the POHR structure is insufficient for fully orienting care toward the actual problem as it exists in the unique patient. Before discussing that crucial point, however, let's discuss another example, this time involving treatment rather than diagnostic decision making.

Alternative treatment options are like alternative diagnostic hypotheses. Each option will have its own set of pros and cons, which vary from one patient to another. One option may be favored by an "evidence-based" guideline derived from population averages (and some other guideline may favor a different option). As applied to an individual patient, however, a guideline's favoring of a treatment option constitutes a hypothesis that the favored treatment is in fact the best option for that individual. This hypothesis must be verified.

Verification requires taking into account all treatment options worth considering for the patient, plus careful selection and analysis of detailed patient data bearing on whether the hypothesized best treatment may in fact be inferior to some alternative treatment once patient-level details are taken into account—details that may be missing (indeed, systematically excluded) from clinical trial data on which the guideline is based. Again, the pros and cons of each treatment option vary from one patient to another. Thus, decision making must methodically work through the details of how that patient would be affected by each option relative to other options, *with the patient being informed and involved in the process*. In this way, decision making becomes oriented toward the actual problem as it uniquely exists in the patient, rather than being oriented toward the abstract problem as it exists in treatment guidelines or in the clinician's mind.

In any event, guidelines are not even established for many issues. Such a guidance vacuum was a root cause of the opioid crisis. When guidelines do exist, they fail to serve their intended purpose for the reasons discussed in Section 3.2.1.

9.1.4 "UNDERSTANDABLE PERHAPS, BUT IT ALMOST KILLED HIM": A FURTHER EXAMPLE

Recall from Section 3.1 the study we cited concluding that most diagnostic errors "were related to patient-practitioner clinical encounter-related processes. … [Improvement] must focus on … factors … that influence the effectiveness of data gathering and synthesis in the patient-practitioner encounter." This conclusion applies to not only diagnosis but the total processes of care.

In her column on diagnosis, Dr. Lisa Sanders describes an emergency suffered by one of her own patients.[225] He was a 64-year-old man who had not been to a doctor for decades. After suffering a stroke, he became a patient of Dr. Sanders. A year passed as she worked on his various medical issues. Then he was rushed to a hospital emergency department with weakness, belly pain,

---

[225] Sanders, L., Missed signals. *The NYT Mag.*, April 27, 2007. Reprinted in *Diagnosis: Solving the Most Baffling Medical Mysteries* (Note 63), pp. 282-286.

an extremely low heart rate, and blood pressure too low to measure. Several steps of diagnostic investigation revealed that his urethra was blocked, which caused his bladder to be distended with four times the normal amount of urine. That pressure caused his kidney to shut down, which caused his cardiac failure. So the diagnostic question became—what caused the blocked urethra?

In a 64-year-old male, a common cause of such blockage is an enlarged prostate. When Dr. Sanders learned what the hospital had determined on her patient, "I felt as if I had been punched in the throat. This was something I should have caught and didn't." But her case involved quite understandable lapses by both herself and her patient. She goes on to candidly explain the clinical and human dimensions of what happened.

> This patient, with his high blood pressure, high cholesterol and stroke, would be at risk for a heart attack, another stroke and, like many men his age, prostate problems. I should have asked about these at every visit and once a year done a rectal exam to assess prostate size and look for cancer. *From reviewing the patient's chart, it appeared I had limited my attention and my exam to his immediate problems—overlooking some of the other risks he faced.*

> I had asked him if he had problems urinating, and he had said no. I don't think he was lying; I think he assumed that his bathroom difficulty was just one more skill stolen from him by his stroke. So much of the damage from that cerebral vascular accident was clearly visible and public. I suspect he felt that this disability, at least, could remain private.

> And when *he didn't acknowledge any difficulties*, I was happy to allow our visits to focus on getting his blood pressure and cholesterol under control, educating him on his medical problems, managing his meds and arranging his transportation and rehab. *Everything else I treated as a long-term goal, to be attended to once these very pressing short-term needs were managed. Understandable perhaps, but it almost killed him.* Practicing medicine is a balancing act—weighing immediate and long-term good. His case was a vivid reminder of what can happen when that balance is lost. [Emphasis added]

This case sheds light on the need for careful, systematic processes, defined by a problem-oriented system of care. Specifically,

- Dr. Sanders focused on her patient's "immediate problems—overlooking some of the other risks he faced." Following a tool-driven, problem-oriented system, in contrast, would mean systematically taking into account prostate issues (not just cancer risk) faced by an older, male patient recovering from a stroke. For example, were CDS tools built for two different problems (urination difficulty and stroke recovery), each tool would reveal the key issue: "Urinary retention in post-stroke patients is significantly related to the poor functional status at initial stage of rehabilitation, and also to poor recovery after rehabilitation." In that light, the mere possibility of urinary retention would be recognized as one of this patient's "immediate problems."[226] As such, the issue would at least be addressed in an initial care plan for recovery from the stroke problem. Better yet, the urinary issue could be included on the problem list as requiring separate attention for diagnosis and management. The care plan for that problem would take into account the stroke recovery issue, which could also be on the problem list.

- The patient himself contributed to his predicament by not acknowledging urination difficulties. That suggests he needed to be educated on his age-related risk of an enlarged prostate and its consequences, which existed regardless of his stroke. "The best way to [educate the patient] is to have the patient … be a full partner in the development and maintenance of the medical record," as LLW recognized decades ago. "In this way the patient will know what data are being gotten and why."[227] Such patient involvement is central to the problem-oriented approach.

- A problem-oriented system is a vehicle for involving not only the patient but non-doctor clinicians and other staff in an organized way. The time pressures on Dr. Sanders indeed made it "understandable perhaps" that she overlooked her patient's prostate risk. Someone else without the same time pressures might have recognized that the patient's negative answer to the question about urination difficulty should be treated as uncertain (given his age alone, not to mention his stroke status). So someone else might have followed up with further inquiry and educating the patient. The point is that such follow-up did not require the scarce and costly time of Dr. Sanders herself. Other personnel in her office could have been mobilized to pursue these activities. A problem-oriented approach facilitates such a division of labor. This is a point emphasized by practitioners who have actually used a POHR in primary care.[228]

---

[226] Son, S. et al., Relation of urinary retention and functional recovery in stroke patients during rehabilitation program. *Ann. Rehabil. Med.*, April 2017; 41(2): 204–210. DOI: 10.5535/arm.2017.41.2.204.

[227] Weed, L. L. et al., *Knowledge Coupling* (Note 18), p. 109. See generally *Your Health Care and How to Manage It* (privately published, 1975, copies available from ldweed424@gmail.com). This point is discussed in detail by Ken Bartholomew's chapter (Note 99) in that book, e.g., at pages 260–63, 274–76. Among other things, he discusses how good records improve communication with patients who appear to have psychosomatic illness.

[228] For further discussion, see Burger, C., The use of problem-knowledge couplers in a primary care practice. *Perm. J.* Spring 2010;14(1):47–50 (DOI: 10.7812/tpp/09-115), and Bartholomew, "The Perspective of a Practitioner" (Note 99), pp. 254, 257–58, 260–61. For example, Ken Bartholomew describes how a medical records technician grew "from being a typist sitting in a cubicle typing whatever I spewed into a dictaphone to a real partner in patient care sitting in my inner office."

- There will always be "very pressing short-term needs" that tend to divert attention from long-term goals. This perpetual "balancing act" demands a technical and administrative infrastructure for coping with it. The mere availability of that infrastructure offers no guarantee that anyone will take advantage of it. But the infrastructure does increase the feasibility of both internal process improvement and effective external scrutiny by outside parties. Lack of the infrastructure, as this case illustrates, "fosters delayed maintenance and crisis care."[229]

9.2 MOVING FROM THE "SCATTERED MODEL" TO A CONSOLIDATED RECORD IN THE CLOUD

Recall from Section 3.4.3 the "scattered model"—multiple records for a single patient scattered among multiple providers. Even if all of those records were to conform to the POHR standard, their scattered state would cause increased risks and inefficiencies. To remedy this problem, LLW long envisioned "a single, integrated, electronic medical record for each person," with all clinicians and the patient feeding their inputs into that record.[230] Cloud-based repositories and tools offer a powerful infrastructure for this "one-patient-one-record" model. That model has been advocated by the Health Record Banking Alliance (HRBA) for a number of years, and an HRBA committee is actively pursuing this model. See the public folder on the committee's Google drive.

To be readily usable, the single record should not be a mere compilation of copies. Instead, the single record should be consolidated, structured, and maintained in accordance with generally accepted standards for problem-oriented records maintained by or for the patient. This model could take one of two basic forms.

- All providers and the patient would jointly use this single record in place of the providers' former separate records. (This model might be called a "unitary health record" or UHR.) This model may require legislative and regulatory changes at state and federal levels, in order to satisfy existing requirements for an authoritative "legal record."[231] Because this UHR would be used by providers as their record system for participating

---

[229] Weed, L. L. et al., *Knowledge Coupling* (Note 18), p. 110.

[230] *Medicine in Denial*, p. 150. This point in the book goes on to cite 2010 testimony by Dr. Jeffrey Schnipper:
...what we need is a single source of truth, that is, one medical record, accessible to providers with permission, and owned by the patient. Otherwise, we perpetuate electronically what we currently have on paper: multiple medical records, each one providing only part of the story. ... In some countries in the developing world, patients bring their chart to every office visit. ... [This] actually solves several problems we have yet to solve: there is one source of truth, there is health information exchange, and it is clear that patients own and are responsible for their medical information.

[231] See American Health Information Management Association (AHIMA), Fundamentals of the Legal Health Record and Designated Record Set.

patients, it would not be solely patient-controlled to the extent that the alternative below would be.

- Patients would be enabled to aggregate copies of their various provider records into a single patient-controlled record. (This model might be called an aggregated health record or AHR.) The patient could allow any or all providers to access that AHR, enabling them to communicate and coordinate with each other and the patient, but the providers would still maintain their own records on the patient. Ongoing updates from the separate provider record systems to the AHR would be needed.

The following uses "personal health record" (PHR) as an umbrella term to cover both the UHR and AHR models and any variations involving a patient-controlled single record.

These PHR models raise a host of policy and technical issues. A full discussion is beyond the scope of this book. The following briefly discusses some of the issues and background.

An important part of the background is that, after considerable evolution, the health IT industry has settled on a technical standard for interoperable exchanges of patient data. The standard is known as Fast Healthcare Interoperability Resources (FHIR®, pronounced "fire").[232] The basic concept of FHIR is to establish standard data elements and formats (known as resources) together with standard application programming interfaces (APIs) for systems to request and respond to requests for transmittal of these data elements. FHIR can be used to build a PHR by obtaining data elements from existing provider records on the patient.

The question is whether that PHR can be problem-oriented. There are currently no generally accepted standards for problem-oriented records, regardless of whether the records are patient-controlled PHRs or traditional provider records. But the private sector (including standards-setting organizations like HL7) has an opportunity to agree on problem-oriented record standards incorporating the data categories outlined in Section 9.1.2 for purposes of building problem-oriented tools. Versions of these tools should be designed for use in both separate provider-controlled record systems and PHR systems for a single patient.

One goal would be to automate, as much as possible, the processes of periodically transmitting the contents of the provider records to the patient's PHR repository (using FHIR), consolidating them into a single PHR, and filtering out old data with no relevance to current or future care. (Further FHIR development would need to take into account the POHR data categories, which are not fully included in FHIR resources.) To the extent these processes can be reliably automated, the need for manual curation of the consolidated PHR would be minimized.

---

[232] The sponsoring organization for FHIR is Health Level Seven International (HL7), a "standards developing organization …for the exchange, integration, sharing, and retrieval of electronic health information …."

The above efforts could eventually become the basis for a public-private initiative to establish generally accepted standards for problem-oriented EHRs. Any such initiative would likely need to include some federal health agencies and other federal entities.[233]

Existing provider EHRs do not implement problem-oriented concepts except in limited, fragmentary and variable ways. These records could still be transferred to PHR repositories, but consolidating them into a single problem-oriented record would be difficult, undoubtedly requiring much manual curation. Experiments on this consolidation need to be undertaken. Depending on the outcomes of these experiments, it may or may not turn out to be feasible and cost-effective on a large scale to consolidate existing records in a usable, problem-oriented PHR. If not, problem-oriented PHRs would be mainly prospective, i.e., excluding the portions of old records unlikely to be relevant for current or future use.

Many advocate a common patient data model (CPDM) "to serve as the reference point for constructing and aggregating all patient data into an optimal representation for end-users … stored in a 'patient cloud'." External EHRs would access and synchronize with the patient cloud using a standard interface (a "clinical portal").[234] This repository would be patient-controlled and would have a patient portal distinct from the clinical portal. The patient portal would provide knowledge to help patients interpret the data (a form of CDS).

Related to the CPDM concept is a federal government effort on common data models (CDMs) for organizing patient data into a common data structure for research purposes. The Office of National Coordinator for Health Information Technology (ONC) has led a project for Common Data Model Harmonization. The project's goal "was to harmonize four CDMs to enable researchers to have access to data from a larger network of patients." This involved building data infrastructure for conducting patient-centered outcomes research using observational data derived from routine health care delivery.

All of the above needs to be examined from the perspective of building a true problem-oriented system of care. From that perspective, there should not be separate repositories with separate data models for patient care and research purposes. Instead, the raw material for clinical research into patient care should be problem-oriented records, generated by a problem-oriented system of care that integrates CDS informational guidance with EHR process guidance. By using that unified CDS-EHR platform for patient-centered care, patients and clinicians would generate EHRs with

---

[233] Largely but not entirely within the Department of Health and Human Services (HHS), the relevant agencies are the Office of the National Coordinator for Health Information Technology (ONC), the National Library of Medicine (NLM) within the National Institutes for Health (NIH), the National Institute for Standards and Technology (NIST), the Center for Disease Control (CDC), the Food and Drug Administration (FDA), the Agency for Health Research and Quality (AHRQ), the HHS Office for Civil Rights (OCR), the Veterans Health Administration (VHA), the Military Health System (MHS), the Innovation Center in the Centers for Medicare & Medicaid Services (CMS), and the Patient-Centered Outcomes Research Institute (PCORI).

[234] See the HL7 Reducing Clinical Burden project white paper, Reducing Clinician Burden by Improving Electronic Health Record Usability and Support for Clinical Workflow (January 2020, p. 17).

scientifically rigorous data in a patient-centered clinical context. Those EHRs would be a uniquely rich and rigorous source of data for research of various kinds, including basic science, clinical care, health services delivery, and health economics.

The following section describes an important effort to conceive and develop such a platform, which we compare with the knowledge coupling and POHR tools previously developed by LLW.

9.3   AKĒLEX: A PLATFORM FOR INTEGRATED CDS AND PROBLEM-ORIENTED RECORD TOOLS

A consolidated PHR needs to ingest, curate, organize, record, and integrate data from various sources, including provider EHRs, CDS tools, wearable devices, pharmacy systems, and more. Managing such a complex data ecosystem will require a new approach—a core knowledge operating system that includes a high degree of CDS and EHR automation. An important effort in this regard has been undertaken by AkēLex (founded in 2001 by Dr. Steve Datena[235]), working with Prosumer Health (founded in 2015 by George Reigeluth as a spin-off from AkēLex), and other partners. AkēLex is building a platform with a unique combination of capabilities. These include sophisticated CDS and EHR functionality that enables a problem-oriented view of normalized patient data for purposes of not only patient care but medical knowledge development. Any party interested in pursuing the ideas in this book should examine what AkēLex is building.

The AkēLex system has four basic components:

- a flexible data ontology (the Lexicon) used to provide an optimized representation of medical knowledge;

- a reference data management system for efficient curation and packaging of medical knowledge for use by the following component;

- the Adaptive Knowledge Engine (AKE), providing a broad range of services (CDS, triage, messaging, quality metrics, resource management, etc.); and

- the Case Entity Record, which stores patient data with associated medical knowledge in a form optimized for future use by the AKE and presented in a problem-oriented structure for clinician users.

Together these components seamlessly integrate CDS and EHR capabilities.

The Case Entity Record: In the Case Entity framework, any clinical entity (diagnoses, procedures, care pathways, medication regiments, even a patient) can be described as a set of attributes or findings collected over time. In patient encounters, when data are collected about the patient's symptoms, conditions, diseases, procedures, labs, response to treatment pathways and other clinical

---

[235] Dr. Datena has assisted with the drafting of this section.

entities, those data relate not only to the patient's case and problem list but also to the broader population of individuals with similar conditions and to the core definitions of individual disease processes. This set of relationships mean that data entered once can be used in caring for the patient, in better understanding the needs of related populations, and powering research designed to increase knowledge about the disease entities themselves.

None of this can be easily accomplished with the kinds of unstructured clinical notations typical of today's clinical documentation. The result is lack of standardization of the how and why of the critical thinking used to make clinical decisions. That makes it difficult to both evaluate the quality of the decision and any insights that might have broader implications for future decisions.

That difficulty is obviated by providing a means of leveraging structured data inputs in support of decision making and research. Making possible this kind of support was the driving force in Dr. Datena's development of the Lexicon, AKE, and Case Entity framework. This endeavor grew out of his experience with systems approaches to care at a level-one trauma center, as described at the History page on the AkēLex website. Like LLW, he saw how clinicians needed external tools to mobilize relevant knowledge and properly capture patient realities in medical records.

The Lexicon: The Lexicon is a data structure allowing for a granular description of any kind of clinical situation or combination of situations in a normalized manner. Structured data can come from various sources (such as clinical histories, exam findings, labs, wearable device data, or CDS assessments), all of which are managed in normalized formats (including consistent terminology), and stored in an EHR with a problem-oriented view. Those data are more granular, accurate, complete, and usable than the raw data found in EHRs and other sources.

The data curation and medical recordkeeping enabled by the Lexicon result in standardized descriptions of clinical entities (medical conditions, procedures, methods, treatment pathways, syndromes, situations, etc.) that the AKE recognizes. The Lexicon thus provides the descriptive language used by the system to articulate queries that the AKE can understand, defines the characteristic data sets relevant to clinical entities managed by the system, and forms the basis of patient records. These capabilities improve decision making and recordkeeping for purposes of not only patient care but knowledge development and process improvement.

The Lexicon structure makes the Case Entity Record created for a patient inherently problem-oriented. The Case Entity Record is a set of time-stamped Lexicon-defined attributes collected through findings from patient encounters over time. Those attributes are stored not only in patient records but in relevant clinical entity datasets defined by the Lexicon. A problem-oriented patient view is simply the *overlap* of the patient dataset (record) with the datasets defining each of the problems (such as diagnoses and treatment needs) assigned to the patient. A problem list is simply a list of the titles of the relevant portion of each of those datasets.

Note that the clinical entity datasets are interconnected. For example, a symptom entity of severe fatigue is connected to its many possible diagnoses, one of which is the Addison's disease entity, which in turn is connected to the established treatment entity (hormone replacement therapy) for that disease. Because the attributes are integral to each and every relevant entity, the framing bias or context of how the data were gathered becomes irrelevant. If an attribute has relevance to a case entity it is applied to all relevant case entities, resulting in a more holistic view of how each attribute fits into the broader context of a patient.

Areas of overlap between the patient and entity datasets define the historical and current status of the patient with regards to each of the entities. Patterns within the overlap may identify the point in the course of the disease or care continuum occupied by the particular patient at a given time (for example, Addison's disease patients vary in what symptoms they experience and when). Associated data exterior to the overlap, but within the definition of a condition case entity, can be used to predict future clinical behavior or drive customized active queries designed to further refine knowledge of the disease state in the patient, or detect adverse patterns early when intervention is most effective. Patient data exterior to the overlap also serve as useful inputs to deep learning models designed to continuously refine the disease entity-associated data sets themselves.

The Adaptive Knowledge Engine:[236] The AKE is a suite of analytics tools for processing normalized data from the Lexicon. The AKE might be said to normalize raw human analytics just as the Lexicon normalizes raw patient data, although "normalize" has different meanings in those two contexts. This decision support by the AKE is provided in multiple contexts (acute diagnostic, therapeutic, chronic disease management, team triage, prevention, and wellness, etc.).

For example, when a patient presents with an undiagnosed complaint, the AKE iteratively uses knowledge from the literature to recommend initial history, physical exam and lab data to collect about the problem. The Lexicon assures that the collected data are entered in the EHR in a normalized form. Then the AKE uses knowledge from the medical literature to perform an "assessment" of the collected data. In the classic CDS use case, the output might take the form of a weighted differential and, if needed, "refinement" questions to aid in further resolution of the query. Once the AKE determines that the data are sufficient to reach a preliminary diagnostic conclusion, the AKE returns a "presumptive" diagnosis. At that point the clinician reviews the record to determine whether the presumptive diagnosis can be accepted as "confirmed." If not, then the clinician proceeds with further investigation using any further guidance that the AKE may provide.

Full transparency in how the system has conducted its assessment is key for developing confidence in the system. Throughout the above processes, the clinician or other user can query the AkēLex system to see exactly what clinical information the AKE is using to make its recommendations, how they are weighted and why. This transparency distinguishes the AkēLex system

---

[236] The company views the AKE's capabilities as a form of AI. We have avoided applying the AI label to the AkēLex system, in accordance with our usage of the AI label in Section 3.4. We believe a virtue of the AkēLex system is that it is closer to traditional software, which is rules-based and transparent in its operations, than to the forms of machine learning employed by advanced AI. See Section 3.3.4.4.

from "black box" advanced AI systems. (Similarly, LLW's knowledge coupling tools as described in Chapter 8 display references to the literature for the options and findings presented.)

AKE tools adjust their output and functionality depending on each user's level of training and background. Thus, doctor, nurse, and patient users might see different information or different displays of the same information.

Comparing the AkēLex system with LLW's tools: The following are points of comparison between the AkēLex system and LLW's tools for knowledge coupling CDS and a problem-oriented EHR. (The comparisons assume a diagnostic scenario, but the comparisons would also apply to a therapeutic decision making scenario.)

- The AKE uses an iterative method of data collection—initial questions, followed by "refinement" questions dependent on the initial answers. The AKE can modify its recommended questions based on use case. For example, triage considerations may require a different level of detail compared to classic CDS. In contrast, the knowledge coupling approach basically has one level. Data collection largely happen all at once, up front, after which the collected data are coupled with applicable knowledge and output is generated.

- Once the AKE analyzes sufficient data, the AKE output may take the form of a recommended presumptive diagnosis (assuming the above diagnostic scenario). In contrast, the knowledge coupling output avoids recommending any one diagnostic option. Instead, it presents a list of diagnostic options worth considering, plus the positive, negative and uncertain findings on each option, plus explanatory knowledge useful for assessing the options and findings in context, with the options grouped in ways relevant to the context and (within each category) ranked by the number of positive findings on each option. See Section 8.3.3.6.

- The informational guidance provided by the AkēLex system and LLW's knowledge coupling tools is quite different in terms of output as the previous bullet explains, but conceptually they are quite similar in the following sense. Just as LLW's combinatorial approach defines a diagnostic option as a combination (finding set) of specific items to be determined from the patient history, physical exam and lab tests, so the AkēLex Lexicon defines a diagnostic entity as a set of a set of Lexicon-derived characteristics. Both systems capture the degree of overlap between the system definition of a diagnostic entity and the real-world finding sets on an actual patient for whom that diagnosis is worth considering. That overlap is usually partial, not complete, reflecting the enormous variability of real-world patients in all their uniqueness—a reality that both systems capture rather than cover up (unlike the generalizations of medical knowledge).

- The AkēLex system integrates decision support with the EHR capability in a way that LLW's knowledge coupling and POHR tools did not. That integration occurs because the Case Entity Record enables Lexicon normalized data to be stored and used for both recordkeeping and knowledge development purposes. Most critically, the stored data can be leveraged by the AKE in all subsequent encounters creating an ever increasing personalization of future assessments and detection of problems only apparent over time. LLW had a similar goal, but his company became focused on knowledge coupling tools and did not pursue integration of the problem-oriented EHR that it first developed in the 1980s.[237]

- Although a problem-oriented view is inherent in the Case Entity Record as explained above, the Case Entity Record in its current form does not include all of the POHR's detailed data categories laid out in Section 9.1.2.2. This means that the process guidance offered by the Case Entity Record differs somewhat from the process guidance offered by the POHR.

- The knowledge coupling tools display the same information to all users, whether doctor, nurse, other clinician, the patient, or a third party. This gives all users the same access to relevant information, and thus protects against dependence on doctors. The AkēLex system stores one master set of data but may vary what is displayed to different classes of users based upon their clinical role and sophistication. This variability in what is displayed enhances ease of use and also facilitates role-based access for HIPAA compliance and other purposes.

---

[237] However, Dr. Ken Bartholomew used early versions of both the knowledge coupling and EHR tools together in his primary care practice. See the detailed description of his experience with both tools in his book chapter cited at Note 99.

CHAPTER 10

World 3 Medicine: Revisiting the Doctor's Role

10.1 MEDICAL TRAINING AND THE DIVISION OF LABOR

As Bates and Gawande wrote about other elements of the health care system, the doctor's role was "never designed with human limitations in mind," but was "rather … built with a series of makeshift patches."[238] Yet, the culture of medicine remains in denial of this reality. The doctor's role, including its authority over other practitioners, is a makeshift social arrangement, a historical accident, not inherent in the needs of modern scientific practice. Yet, the doctor's role so ingrained that it seems linguistically incorrect to speak of medical practice by a nurse, or to equate medical and nursing practice.

Breaking through this state of denial requires demonstrating a viable alternative to the doctor's role. Only seeing and experiencing that alternative will fully expose why the current role is not workable and not needed.

A viable alternative requires a new informational supply chain based on new tools for managing health information, as set out in Chapters 7–9. A viable alternative also requires a new division of labor, with new concepts of training and licensure for medical practitioners, as discussed in part VIII of *Medicine in Denial* (Note 4). The following explains and expands on portions of the analysis there.[239]

The doctor's current role evolved out of Abraham Flexner's famous 1910 report on medical education.[240] At that time, many doctors were educated outside of universities. They attended trade schools with low admissions standards, and much of their learning occurred through apprenticeships. Their training largely did not incorporate scientific advances. Rejecting this model, Flexner advocated the Johns Hopkins, post-graduate model of education, founded in basic science, conducted at universities, and oriented toward research more than practice.

---

[238] *Error in Medicine: What Have We Learned?* (see Note 32).

[239] In addition, see our 2014 article, "Diagnosing diagnostic failure" (Note 20).

[240] Flexner, A. (1910), Medical Education in the United States and Canada: A Report to the Carnegie Foundation for the Advancement of Teaching. Bulletin No. 4., http://archive.carnegiefoundation.org/publications/pdfs/elibrary/Carnegie_Flexner_Report.pdf. New York City: The Carnegie Foundation for the Advancement of Teaching.

Flexner's model was seen as the only way to bring scientific advances to medical practice. These were viewed as advances in knowledge—which overlooked the advances in intellectual behavior that engendered modern science, as Francis Bacon first envisioned. And knowledge was seen as residing in the human mind (Karl Popper's World 2), rather than as objective content existing independently of the mind (World 3), as discussed in Chapter 6. Flexner thus missed the crucial insight of his contemporary, Whitehead, who saw that civilization advances by lessening dependence on human thought—an insight that Hayek applied to the domain of commerce, a domain that overlaps with medicine.[241]

The outcome of Flexner-inspired reforms was that medical education was expected to instill a core of scientific knowledge. Neglected was the need for a core of scientific behavior. Applying knowledge with scientific rigor is a behavioral skill to be learned. But learning that skill should not be equated with learning knowledge in the abstract. And such didactic learning should not be a basis for licensure.

The notion of instilling a core of scientific knowledge turns out to be both futile and unnecessary. Its futility is that medical knowledge is constantly changing and expanding, so that no human mind can keep up with it. And keeping up with medical knowledge by *learning* (Popper's World 2) is unnecessary. because medical knowledge is a World 3 resource, now more available and usable than ever before with electronic tools.

From this point of view, consider what Dr. Siddhartha Mukherjee has to say about his medical education:

> When I began medical school in the fall of 1995, the curriculum seemed perfectly congruent to the requirements of the discipline. I studied cell biology, anatomy, physiology, pathology, pharmacology. By the end of four years, I could recognize the five branches of the facial nerve, the chemical reactions that metabolize proteins in cells, and parts of the human body that I did not even know I possessed. … *But all this information could, I soon realized, be looked up in a book or found by a single click on the Web.*[242] [Emphasis added.]

This realization comes to many doctors. Marty Makary writes that "most medical schools still haze students by making them commit to memory thousands of details that do not need to be rapidly recalled in the real world of doctoring."[243] Moreover, some of those thousands of details, if not simply forgotten, are destined to be proven wrong or otherwise become obsolete. Thus the saying told to new medical students—"half of what you will learn is wrong, but we don't know which half."[244] In addition, medical students are expected to memorize minutiae that do not become

---

[241]   See Chapter 6 and *Medicine in Denial* (see Note 4), part V at pp. 107, 122.
[242]   *The Laws of Medicine: Field Notes from an Uncertain Science*, (Simon & Schuster, TED Books, 2015), p. 5.
[243]   *The Price We Pay* (Note 53), p. 231.
[244]   See Variations on the "but we don't know which half" line.

obsolete but also never become usable in practice. An example often mentioned by LLW is learning the Krebs cycle. Of this example Dr. Makary writes:

> During my medical education, a dozen different times I had to memorize the Krebs cycle, a series of names of changing molecules inside a cell. I took a written exam almost every year to see if I could quickly recall the names of the intermediate molecules in the Krebs cycle. Of course, I could have used my brain space for things relevant to my patients, and if not, I can always look up the names of these molecules. The Krebs cycle has not come up once in my years of clinical medicine in any way, shape, or form.[245]

Of course, doctor training involves much more than such rote learning. But that "much more" is needed precisely because all the rote learning accomplishes so little.

So, can we at least hope that the "much more" ultimately does assure competent performance by doctors? We have no such assurance. On the contrary, doctor training is notorious for neglecting core competencies:

> Core bedside skills of history taking and physical examination—still vital to comprehensive assessment, diagnostic accuracy, and truly patient-focused care—are taught and assessed in the first two years of medical school but largely ignored once the student reaches the clinical years. During residency, development of these skills is assumed when in fact they wither further. The physical examination of newly admitted patients is often cursory and, what is worse, perverted by drop-down boxes into an exaggerated and invented form that reads better than the truth. [246]

In addition to these clinical skills, doctors also need behavioral skills. Dr. Makary writes of his lengthy training:

> Noticeably absent from that 15 years of study were the behavior skills that enable doctors to perform well. … I learned the Krebs cycle, but not how to communicate effectively with nurses. I learned the microscopic stages of prostate cancer, but not how to deal with an underperforming person on my surgical team. I learned subatomic particles for the MCAT, but never learned how to explain diabetes at a sixth grade reading level. To address the serious gaps in education, the traditional 15-year track to becoming a specialist doctor needs one giant enema.[247]

---

[245]   *The Price We Pay* (Note 53)), p. 230.
[246]   Elder, A. et al., *The Road Back to the Bedside*. 2013;310(8):799–800. See also Dr. Abraham Verghese's rich and vivid description of the manifold benefits—medical, economic, and psychic—of skillfully conducting physical exams for diagnosis, in his article, Culture shock—Patient as icon, icon as patient. *N. Engl. J. Med.* 2008;359;26:2748–2751, https://medicine.stanford.edu/content/dam/sm/medicine/documents/education/bedsiderevisited/Verghese.pdf.
[247]   *The Price We Pay* (Note 53)), p. 230.

Even if one accepts learning medical knowledge as a goal of medical education and even if there were some identifiable, unchanging core of knowledge needed by all doctors, medical school curricula fail to cover some crucial subjects.

A prominent example is nutrition. This was a known gap in medical education in 1985, when a report from the National Academy of Sciences recommended a minimum of 25 curricular hours devoted to nutrition in medical school. Yet, since then actual curricular hours on nutrition have not met this minimum and have actually declined.[248] A 2019 report concludes, "despite the overwhelming evidence linking food with health, nutrition receives little attention in medical school and throughout the education of physicians."[249] This report recommends increasing coverage of nutrition in medical school curricula, adding nutrition coverage to medical school accreditation requirements, tying federal funds for medical schools to nutritional education, and adding nutrition-focused content to standardized exams relied on for residency admissions, licensure, board certification, and continuing education.

Such recommendations unthinkingly accept the traditional paradigm—an informational supply chain dependent on the personal knowledge of highly educated doctors. Nutrition is but one subject matter example of the futility of this paradigm. The 2019 report acknowledges that medical school curricula and standardized tests "are already very crowded" with other subject matter. Moreover, testing changes alone "may not improve physicians' general knowledge of nutrition or their ability to counsel patients" (p. 24).

These difficulties illustrate that doctors have long been overburdened with knowing and applying more knowledge than they can handle. This burden is especially pointless for nutrition because that field already has a large supply of non-doctor specialists.[250]

In situations like that, involving a common body of knowledge such as nutrition, what exactly are the licensure boundaries between doctor and non-doctor practitioners? Another situation like that is illustrated by "the case of the crazy surgeon and the wise physical therapist" described in Section 1.1.1. One of the two physical therapists involved, not the other one and certainly not the orthopedic surgeon, had the relevant expertise, and that expertise evidently came from experience, not training.

Issues like these lead to various questions. What is the point of knowledge-based medical education and licensure for doctors and other health professionals? As to doctors, how can their

---

[248]   How much does your doctor actually know about nutrition? *Am. Heart Asso. News.* (May 3, 2018), discussing a report "in the journal *Circulation* that looked at gaps in nutrition education over the decades."

[249]   Food Law Clinic, Harvard Law School, Doctoring our diet: Policy tools to include nutrition in U.S. Medical training (September 2019), p. i.

[250]   See the *Occupational Outlook Handbook* on Dieticians and Nutritionists (74,200 positions in 2019). There are also health educators and community health workers in greater numbers, who are often more available than other professionals, and who could be equipped with CDS tools to help patients with nutrition issues. See the *Occupational Outlook Handbook* on *Health Educators and Community Health Workers*. (127,100 positions in 2019). For materials on the importance of community health workers, see https://chw.upenn.edu/.

training possibly justify all the time it takes, the high costs it imposes, and the monopolistic authority it confers? What is the alternative to knowledge-based education and licensure? And if there is a better alternative for doctors, should it also apply for other health professionals? For that matter, why should the occupational category of "doctor" continue to exist as a separate category? Why should doctors' knowledge-based training, with all its failings, give them a legal monopoly over medical practice and placement at the top of the medical hierarchy? After all, doctors don't monopolize the interpersonal and manual skills involved in medical practice.

These questions are especially acute for primary care. In that arena, the doctor's traditional role is a straitjacket, limiting the availability and flexibility of the health care workforce. This dilemma has led many to envision alternative roles, such as the following:

> What we need are primary care extenders with local ties and cultural competence of community health care workers, the procedural skills of PAs, and ready access to the knowledge of NPs and primary care physicians. They should be easy to train, inexpensive to employ, and capable of working miles apart from their supervising providers. This model does not exist in primary care, but it does in a related field: emergency medical services (EMS).
>
> EMS is almost entirely delivered by emergency medical technicians (EMTs) and paramedics. EMTs extricate people from car crashes, control bleeding, splint fractures, and provide basic life support. Paramedics conduct detailed patient assessments, insert intravenous lines, administer a wide range of oral and parenteral drugs, and perform certain lifesaving procedures.
>
> If EMS professionals can deliver lifesaving care miles from their supervising physicians, shouldn't comparably trained extenders—primary care technicians (PCTs)—be used to improve access to primary care? The idea isn't new; it's been described since the 1970s.[251]

Community health workers[252] would be obvious candidates to be trained as PCTs.

Questions about the utility of the doctor's traditional role (and corresponding questions about medical training and licensure) are still outside mainstream health policy debates. But such questions are implicit in many critiques. A recent example is from Dr. Ezekiel Emanuel and Emily Gudbranson: "By prioritizing academic pedigree, the medical profession has traditionally overemphasized general intelligence and underemphasized—if not totally ignored—emotional intelligence."[253] They further argue that general intelligence as shown in academic settings may not

---

[251] Kellerman, A., et al., Primary care technicians: a solution to the primary care workforce gap. *Health Affairs*, 2013;32(11):1893–1898. As of 2019, the number of jobs classified as EMTs and Paramedics is 265,200, https://www.healthaffairs.org/doi/full/10.1377/hlthaff.2013.0481.

[252] See Note 250.

[253] Emanuel, E., Does medicine overemphasize IQ? *JAMA* Vol. 319, No. 7 (February 20, 2018).

correlate with real-world performance, intellectually and otherwise. "For instance, good test takers can score extremely high on multiple-choice examinations but may lack real analytic ability, problem-solving skills, and common sense."

These points are headed in the right direction but stop short of where we need to go. Why not rethink established occupational roles, for example, the roles of nurses? Relative to doctors, could it be that nurses tend to be superior in emotional intelligence and comparable in "real analytic ability, problem-solving skills, and common sense"? Why are doctors higher than nurses and other clinicians in medicine's hierarchy? Given that so much expense and professional time is (or should be) invested in activities such as "sensitively discussing end-of-life care preferences with a patient who has developed metastatic cancer," or "effectively help[ing] patients change their behaviors," could it be that nurses should play the leading role in such activities? Or should such activities be conducted by "health coaches" as a distinct category of medical practitioner? Should doctors just occupy niches requiring specific skills that some of them uniquely possess? If such niches exist, are they just self-fulfilling prophecies imposed by the different educational and licensure tracks for doctors and nurses? Maybe nurses could acquire capabilities now reserved to doctors in specific skills were they permitted to do so and demonstrate their competence. This is long been happening to some extent with physician assistants and nurse practitioners (PAs and NPs).

There is much talk about how both doctors and nurses and others should "practice at the top of their license," without being burdened by lower-level tasks. But exactly what tasks are at "the top of their licenses" and how is this determined? There are valuable innovations in medical school curricula and teaching, such as those described in Dr. Makary's recent book.[254] But those are innovations in doctor training, not disruptions in medical licensure and continued existence of the doctor's role under state law.

These questions become all the more relevant with the emergence of powerful AI tools. "As we think about the role of each and every clinician," Eric Topol writes, "it becomes increasingly clear that AI has a potential transformative impact."[255]

Medicine's culture of denial blocks awareness of the possibility that occupational categories like doctor and nurse are anachronisms. Truly facing that possibility would lead inexorably to the conclusion that doctors' legal monopoly over medical practice is not justified.

Basic questions like these led LLW to conclude that education and licensure for all medical practitioners, not just doctors, should be skills-based rather than knowledge-based, as discussed in part VIII of *Medicine in Denial*. That led him to further conclusions about dividing up the doctor's role into many different practitioner roles, all of them based on mastery of discrete skills, all of them

---

[254] *The Price We Pay* (Note 53), p. 232, describing the curricular innovations of Dr Stephen Klasko, CEO of Thomas Jefferson University and Jefferson Health (Dr. Klasko found that "the way we select and train physicians is akin to joining a cult"). By necessity under state medical practice acts, such innovations do not alter the legal status and authority of doctors.

[255] *Deep Medicine* (see Note 7), p. 182.

using a common infrastructure of external tools, and none of them requiring the extraordinarily prolonged and costly training that doctors endure now.

LLW wrote about these issues 40 years ago, when he was starting to develop "knowledge coupling" CDS tools:

> Since much of the knowledge and medical logic [possessed by doctors] can be available in new communication tools themselves, anyone who demonstrates proficiency in the procedures of medicine … should be employed without having to present credentials based on memorialized knowledge or a capacity to generate complete logic pathways extemporaneously. … Under our present licensing and credential system, which focuses on training rather than performance, throughout one's professional life, there is far too much reward for those doing simple things or doing many complex things poorly. …
>
> In the selection of health providers, natural skills in interpersonal relations, manual dexterity, dedication, and a sense of responsibility to others should far outweigh the present emphasis on memorized knowledge and extensive backgrounds in formal education.[256]

A radically new division of labor must evolve. The doctor's role must be disaggregated into multiple roles performed by medical practitioners of every description. Some practitioners will choose to specialize in only one skill or a narrow range of closely related skills. Others will master a broader range of skills. The scope of their licenses will vary accordingly. No medical practitioners should have the broad licenses and autonomy currently conferred on doctors. Their licenses do not assure that they maintain and update whatever mastery of skills they may develop initially. This lack of assured mastery of execution inputs is one of the ways scientific rigor is lacking in the health care system.

Doctors may find such change in their traditional role deeply threatening. But the status quo leaves them deeply harmed. Their monopolistic credentials burden them with unaffordable education debts, unattainable standards of care, unmanageable third party demands, unpredictable litigation exposure, and unbearable risks of error for their own patients.[257]

10.2   PROFESSIONAL AUTONOMY

The central ideas of this book may be difficult to take seriously for doctors who accept the traditions of their profession. From their perspective, the book's central ideas would dismantle a time-honored tradition of highly educated doctors autonomously using their sophisticated expert

---

[256] *Physicians of the Future* (Note 17), p. 904.
[257] See our 2011 blog post, Is Fee-for-Service Really the Problem?

judgment to apply advanced medical science. All that would be replaced with a scheme of "problem-oriented" care, delivered by seemingly lower-level practitioners than doctors, driven by outside parties through external tools designed to impose one-size-fits-all standards of care.[258] On this view, the detailed, pre-defined data collection imposed by CDS tools as described in Chapter 8, and the detailed data categories for EHRs as described in Chapter 9, are inflexible, one-size-fits-all impositions on the judgments of autonomous practitioners who know best what works for their own patients and practices.

This view is based on a false ideal of professional autonomy. Autonomy should not mean that professionals are each free to set their own rules. Any complex adaptive system depends on basic rules for foundational elements (see Chapter 6). The CDS and EHR tools we have described simply define uniform basic standards for orderly problem solving. Their level of detail simply reflects specificity.

Consider a couple of rough historical analogies, both involving "one-size-fits-all" standards established for good reason. First, as we discussed in Section 6.2, the core concepts of double-entry bookkeeping spread all over the world to become the basis of "generally accepted accounting principles" for managing financial information. The core concepts of problem-orientation can play a similar role for managing clinical information. Second, W. Edwards Deming's theory of management for "manufacturing processes … [and] the processes by which enterprises are led and managed" came to have "enormous influence on the economics of the industrialized world after 1950."[259] Something like that could happen in the health care system. . LLW's ideas offer a pathway to define and enforce high standards of care for quality and continuous improvement, by using external tools for the informational supply chain.[260] What these two analogies suggest is that medicine is historically anomalous in its lack of such generally accepted standards for elemental processes.

The comparison with Edwards Deming is worth examining further. Beginning around 1950, Deming was effectively exiled to Japan, as his work was ignored for decades in the U.S. The Japanese patiently learned and applied Deming's principles. This learning helped give rise to "what has become known as the Japanese post-war economic miracle of 1950 to 1960" (per Wikipedia). The Japanese developed a potent competitive advantage and became the second largest economy in the world. In the meantime, awareness of Deming's work did not even begin to spread in the U.S. until 1980. American industry maintained a state of denial about its decline, until eventually the mortal

---

[258] We do not address here another central idea of this book—the primacy of the patient's role. Most doctors now at least pay lip service to that role, recognizing that traditional paternalism is unworkable. For further discussion, see *Medicine in Denial* (Note 4), pp. 5–8 (analogizing the patient's role to the traveler's role in the transportation system) and part IX (discussing autonomy and education for the patient/consumer, who necessarily acts as one of his or her own providers).

[259] See https://en.wikipedia.org/wiki/W._Edwards_Deming.

[260] A related concept is that of "simple rules" governing everyone in a complex adaptive system. For further comparisons of rules for financial accounting and complex adaptive systems, see *Medicine in Denial* (Note 4) on pp. 34, 91, 121, 127–29, 140–41, 174.

competitive threat from the Japanese broke through the denial. Andrew Grove, the late CEO of Intel, summarized what happened in the 1980–2000 period:

> In the 1980s American manufacturers in industries ranging from automobiles to semiconductors to photocopiers were threatened by a flood of high-quality Japanese goods produced at lower cost. Competing with these products exposed the *inherent weakness in the quality of our own products*. It was a serious threat. At first, American manufacturers responded by inspecting their products more rigorously, putting ever-increasing pressure on their quality assurance organizations. I know this first-hand because this is what we did at Intel.
>
> *Eventually, however, we and other manufacturers realized that if the products were of inherently poor quality, no amount of inspection would turn them into high-quality goods.* After much struggle—hand-wringing, finger-pointing, rationalizing and attempts at damage control—we finally concluded that *the entire system* of designing and manufacturing goods, as well as monitoring the production process, *had to be changed. Quality could only be fixed by addressing the entire cycle*, from design to shipment to the customer. This *rebuilding from top to bottom led to the resurgence* of U.S. manufacturing. [Emphasis added.][261]

Medicine is still in a state of denial, still going through what Andrew Grove described—"hand-wringing, finger-pointing, rationalizing and attempts at damage control." Just as American industry had to face up to the "inherently poor quality" of its products, so medicine now has to face up to the inherently poor quality of care driven by clinicians' fallible judgments and poorly assembled information. Facing up to the problem means "rebuilding from top to bottom" the care delivery system. This will involve deeper disruptions (including change in occupational hierarchies) than those involved in the resurgence described by Andrew Grove.

The decline, rebuilding and resurgence of U.S. manufacturing was followed by a spasm of corporate misdeeds in cases like Enron. Andrew Grove confronted this issue by identifying the root cause and advocating specific reforms: "Corporate misdeeds, like poor quality, are a result of a systemic problem, and a systemic problem requires a systemic solution. I believe *the solutions that are needed all fit under the banner of "separation of powers"* [emphasis added]. Grove criticized corporate governance for its "built-in conflict" of interest. The "systemic solution" he advocated involved checks and balances through separation of corporate oversight powers, including powers over financial accounting.

The health care system faces an analogous predicament in two respects: (a) excessive concentration of powers (and burdens) in doctors, due to their legal and cultural monopoly over

---

[261] From a July 17, 2002 *Washington Post* op-ed, reproduced in the *Congressional Record*. The title of Grove's piece was "Stigmatizing Business," and he criticized simplistic attacks on corporate executives who were struggling to better compete. *But he did not defend the status quo.*

medical practice (with its "built-in conflict" of interest for both fee-for-service and capitated environments[262]); and (b) lack of meaningful standards of care for managing clinical information (by comparison, the corporate world had generally accepted standards for managing financial information but allowed those standards to be corrupted in cases like Enron).

The solution for medicine is similarly twofold: (i) disaggregating the multiple roles performed by doctors into many separate roles performed by many categories of medical practitioners, each licensed to perform specific skills based on demonstrated competence; and (ii) enforcing high standards of care for managing clinical information, and with corresponding tools used to integrate the functioning of multiple practitioners caring for any patient, using information tools designed for these purposes. These solutions are essential to bridge the gap between the human mind and medicine's information burden. They are also needed to address three related problems: the serious conflict of interest built into the doctor monopoly, lack of transparency and accountability, and hopeless regulatory complexity.[263]

These solutions are also needed to bring more humanity to medical practice. Practitioners must become better attuned to patient experiences with their own disease conditions and their own care—a recurrent issue in this book. See especially the case examples in Section 1.1 and Chapter 3, the "Initial problem definition" elements in Section 9.1.2.2, and the discussion of the "subjective" data category at Notes 214–216 in Section 9.1.2.2. See also Note 211 citing LLW's critique of the misguided "APSO" alternative to SOAP notes.

At the very core of the healer's role is recognizing what patients need to be healed from. However, all too often that recognition is incomplete or absent altogether. "If doctors can't find a biological explanation for what's troubling their patients, patients have trouble being believed," as described in a remarkable new article on "long COVID" patients who endure persistent symptoms.[264] This phenomenon is pervasive:

> Even with growing awareness about long Covid, patients with chronic "medically unexplained" symptoms—that don't correspond to problematic blood tests or imaging—are still too often minimized and dismissed by health professionals. It's a frustrating blind spot in health care, but one that can't be as easily ignored with so many new patients entering this category, said Megan Hosey, assistant professor at the Johns Hopkins Department of Physical Medicine and Rehabilitation. …
>
> "A proportion—usually around 30%—of survivors of any medical condition report high rates of fatigue, sleep disturbance, brain fog, pain, depression, and anxiety that interfere with their ability to live fully" [quoting Dr. Hosey]. Long before the

262   See *Medicine in Denial* (Note 4), p. 35 and our 2011 blog post, Is Fee-for-Service Really the Problem?
263   For further discussion, see our 2011 blog post, Is Fee-for-Service Really the Problem?
264   Belluz, J. Long Covid isn't as unique as we thought, *Vox*, March 11, 2021.

pandemic, the intensive care community coined a term for the persistent symptoms people frequently experience following stays in an ICU for any reason ....

... Researchers who've investigated psychiatric and psychosomatic triggers for chronic conditions like ME/CFS [myalgic encephalomyelitis/chronic fatigue syndrome], for example, haven't found a consistent link. "So maybe that really would not be the first direction you would go in with long Covid. ... These are biological diseases driven by biological causes and they really don't seem to be diseases of the psyche." ...

If long Covid changes anything, it has to be this knee-jerk reaction in medicine to discount and give up on patients with symptoms that have no identifiable biological basis. The experience is so pervasive that researchers at the Mayo Clinic in Minnesota gave it a name: "undercared-for chronic suffering."

... Simply put, medicine hasn't cracked how to deal with patients who have chronic syndromes, like ME/CFS or long Covid, that don't have one-size-fits-all treatment regimens. ...

Instead, long-haulers of any chronic condition can exist in a space between sickness and health for years, sometimes without a diagnosis. Their unexplainable symptoms can elicit skepticism in health professionals, ... who are trained to consider patient feedback the "lowest form of evidence on (the evidence hierarchy), even under research on mice" [quoting Amy Proal, a microbiologist].

With wartime levels of long-Covid patients now flooding health systems around the world, "it's time for medicine to be rooted in just believing the patient," Proal added.

If the authoritative role of doctors permits them to just *disbelieve* the patient, their role leads to not only dehumanizing the care they provide but distorting the scientific data they enter or fail to enter in health records. That distortion undermines scientific research into "biological diseases driven by biological causes" and whatever mind–body connections may be involved.

Changing this state of affairs is inseparable from changing the design and use of information tools on which the roles of doctors, other clinicians, and patients depend. As to the role of the tools, LLW cited an aphorism from Thomas Carlyle: "Man is a tool-using animal. ... Without tools he is nothing; with tools he is all."[265]

Given the legal and cultural authority over medical practice conferred on doctors, they are positioned to take the lead in seeking better tools and using those tools to reform their own role. If they don't, then others will.

---

[265] Sartor Resartus, *Book 21*, Chapter 5 **(1834)**.

CHAPTER 11

Conclusion

As noted in Section 1.1, much of the critique in Part I of this book has become familiar. Now that some of LLW's insights have become conventional wisdom, it is long past time to face up to what we are still missing in LLW's work.

LLW's insights grew out of his pursuing concrete solutions for real-world problems. What we are still missing is continued pursuit of the solutions—the tools and standards and underlying principles described in Chapters 7, 8, and 9, plus LLW's re-conception of occupational roles and licensure in medicine (Chapter 10). Medicine needs more than piecemeal adoption of fragments like problem lists and SOAP notes. It needs to experiment aggressively with working models of a total system of problem-oriented care, including feedback loops and enforcement for accountability and continuous improvement.

Engaging in that pursuit will deepen insight into the problems, as it did for LLW. Most of all, that pursuit will take us farther down the road of reforming the health domain at the deepest level—better medical practice by health professionals and institutions combined with more effective pursuit of health by individuals and communities.

This is not to deny the potential for important improvements within the current paradigm (see Note 15). But those improvements are not enough. Leaders need to pursue redesigning the health care system from its foundation in medical practice and thereby transforming the entire health domain.

People may disagree about where such reform will lead and how much good it will do. But the only way to find out is to pursue such reform step by step. If we shrink from that pursuit because of its threat to the doctor-centric status quo, then our difficulties will remain intractable. If we engage in that pursuit, then the health domain may change from a threat to a hope for the future.

Correction to: Ending Medicine's Chronic Dysfunction: Tools and Standards for Medical Decision Making

Lawrence L. Weed and Lincoln Weed

Correction to:

L. L. Weed and L. Weed, *Ending Medicine's Chronic Dysfunction: Tools and Standards for Medical Decision Making*
https://doi.org/10.1007/978-3-031-01607-3

The author Lawrence L. Weed was not captured in this book. This has now been corrected.

The updated version of this book can be found at
https://doi.org/10.1007/978-3-031-01607-3

L. L. Weed et al., *Ending Medicine's Chronic Dysfunction: Tools and Standards for Medical Decision Making*
© Springer Nature Switzerland AG 2023
https://doi.org/10.1007/978-3-031-01607-3_12

Acknowledgments

This book reflects the contributions of many people over decades. Lincoln Weed wishes to acknowledge the following in particular.

In the 1970s Professor Ron Baecker was asked to examine the work of the PROMIS Laboratory, headed by Lawrence L. Weed (LLW). His memory of meeting with LLW and colleagues led him to contact me more than 40 years later, when he invited me to prepare this book for publication in the Morgan & Claypool book series he edits. I am greatly indebted to him for offering me this opportunity. I am also indebted to the Morgan & Claypool staff (Diane Cerra, Christine Kiilerich, Deb Gabriel) for their expertise and patience with my endless revisions.

Harold Cross, MD, was present at the creation of the problem-oriented record in the late 1950s, when he followed LLW from medical school at Yale to Bangor, Maine. Harold has been a family friend ever since. He pioneered use of the problem-oriented record in his own medical practice and co-authored an influential book on the subject with his late colleague John Bjorn, MD (cited at Note 151). He remains active and provided valuable comments on a draft of this book.

Charlie Burger, MD, was a student of LLW's at the Case Western Reserve medical school in the 1960s, where he became committed to LLW's principles. Later, he joined Harold Cross's practice in the Bangor, Maine area and became a pioneering user of LLW's knowledge coupling tools (see his article cited at Note 228). This work has contributed to his being nationally recognized for innovations in primary care by several leading organizations. He remains active as the Chief Health Officer of Prosumer Health. I am grateful to him for his many efforts to advance LLW's ideas, including comments on a draft of this book.

In 1993, the late Dale G. Breaden, as editor of the *Federation Bulletin: The Journal of Medical Licensure and Discipline*, heard LLW speak to the Federation of State Medical Boards of the United States. He invited LLW to submit an article to the *Federation Bulletin*. That gave me an excuse to start working with LLW on the proposed article, which was published in 1994 as "Reengineering Medicine" (see Note 124). He wrote a thoughtful and generous editorial introducing the article. It was apparent that Mr. Breaden, who was a student of history without medical training, had a broader perspective on LLW's ideas than many doctors.

George Reigeluth, founder and CEO of Prosumer Health, first met LLW in 2005. That began 12 years of dialogue between them. From the beginning he saw LLW's ideas as centrally important, and he became committed to advancing them, which he is continuing to do with Prosumer Health and the HRBA. With his long experience as an entrepreneur and his background in public health and health policy, he brought a fresh perspective and LLW greatly valued his involvement. I am especially grateful to him for getting my brother Chris and me involved with the Health Record Banking Alliance (HRBA).

I am grateful to the members of the HRBA Infrastructure Committee, who understand the relevance of the problem-oriented record to their mission, who have invited extensive input from Chris and me, and two of whom have provided detailed and valuable comments on this book. Moreover, the collective expertise of the committee members has been highly educational for me, and has led to the opportunity to participate in HL7 EHR Workgroup projects.

I am grateful to those who commented on a draft of this book, including but not limited to those mentioned above. None of them have seen the final version, however, and none bear any responsibility for whatever errors or deficiencies remain.

Finally, I cannot begin to convey the debt I owe to my family, including my late mother and my brother Chris (some of their enormous contributions to LLW's work are described at the end of the *Medicine in Denial* book), my late wife Margo Stever, and my daughter Julia (whose contributions to my life include providing a crucial comment on a draft of this book).

Authors' Biographies

Lawrence L. Weed, MD (1923–2017) is best known for originating problem lists and "SOAP notes," two components of the problem-oriented record standard for organizing data in health records. As discussed in Chapter 5, LLW's work on health records arose out of his experiences in medical school, internships, residency in internal medicine, and basic research in biochemistry, at several institutions. In 1965, he became director of the outpatient clinics at Cleveland Metropolitan General Hospital, where he established a group to develop a computerized problem-oriented record. In 1969, he moved to the University of Vermont, where his group became known as the PROMIS Laboratory. LLW received a number of awards for his work, most prominently the Gustav O. Lienhard Award in 1995 from the Institute of Medicine of the National Academy of Sciences. Further details on his career are available in a *New York Times* obituary and other sources cited in chapter 5, Note 129.

Lincoln Weed, JD, a son of LLW, practiced employee benefits law in Washington, D.C. for 26 years, followed by 8 years at a consulting firm where he specialized in health privacy. His experience as an employee benefits lawyer included work on health benefits. This intersected with LLW's work in medicine, which led to them co-authoring several publications. He can be reached at ldweed424@gmail.com and 703-424-4408.

Printed in the United States
by Baker & Taylor Publisher Services